Springer Theses

Recognizing Outstanding Ph.D. Research

Aims and Scope

The series "Springer Theses" brings together a selection of the very best Ph.D. theses from around the world and across the physical sciences. Nominated and endorsed by two recognized specialists, each published volume has been selected for its scientific excellence and the high impact of its contents for the pertinent field of research. For greater accessibility to non-specialists, the published versions include an extended introduction, as well as a foreword by the student's supervisor explaining the special relevance of the work for the field. As a whole, the series will provide a valuable resource both for newcomers to the research fields described, and for other scientists seeking detailed background information on special questions. Finally, it provides an accredited documentation of the valuable contributions made by today's younger generation of scientists.

Theses are accepted into the series by invited nomination only and must fulfill all of the following criteria

- They must be written in good English.
- The topic should fall within the confines of Chemistry, Physics, Earth Sciences, Engineering and related interdisciplinary fields such as Materials, Nanoscience, Chemical Engineering, Complex Systems and Biophysics.
- The work reported in the thesis must represent a significant scientific advance.
- If the thesis includes previously published material, permission to reproduce this must be gained from the respective copyright holder.
- They must have been examined and passed during the 12 months prior to nomination.
- Each thesis should include a foreword by the supervisor outlining the significance of its content.
- The theses should have a clearly defined structure including an introduction accessible to scientists not expert in that particular field.

More information about this series at http://www.springer.com/series/8790

Xu Yang

A Study on Antimicrobial Effects of Nanosilver for Drinking Water Disinfection

Doctoral Thesis accepted by
National University of Singapore, Singapore

Springer

Author
Dr. Xu Yang
Department of Civil and Environmental Engineering, 1 Engineering Drive 2
National University of Singapore
Singapore

Supervisors
Prof. Say Leong Ong
Department of Civil and Environmental Engineering, 1 Engineering Drive 2
National University of Singapore
Singapore

Assoc. Prof. Jiangyong Hu
Department of Civil and Environmental Engineering, 1 Engineering Drive 2
National University of Singapore
Singapore

ISSN 2190-5053 ISSN 2190-5061 (electronic)
Springer Theses
ISBN 978-981-10-2901-1 ISBN 978-981-10-2902-8 (eBook)
DOI 10.1007/978-981-10-2902-8

Library of Congress Control Number: 2016954595

Printed on acid-free paper

This Springer imprint is published by Springer Nature
The registered company is Springer Nature Singapore Pte Ltd.
The registered company address is: 152 Beach Road, #22-06/08 Gateway East, Singapore 189721, Singapore

Supervisor's Foreword

The presence of waterborne pathogen in drinking water has been causing diseases around the globe. Diarrheal disease, for instance, is caused by a host of waterborne bacterial, viral, and parasitic organism, accounting for 1.8 million deaths every year according to WHO report published in 2004. What is worth noting is, even in developed countries, waterborne pathogen is one of the main threats to public health. In Singapore, there were 127,402 attendances reported as acute diarrheal illnesses in 2012. Situations in countries lacking basic water treatment practices are even worse.

Early attempts at implementing disinfection in water treatment process can be dated back to 1890s, when chlorination was first introduced in Germany and England. The importance of chlorination can never be overemphasized for its contribution toward saving countless people from infection over the last century. However, chlorination could be problematic in some circumstances. Chlorine can react with organic compounds, producing disinfection by-products (DBPs) that can be potentially carcinogenic in nature. Subsequently, some alternative technologies, such as ozonation and chloramination, were implemented to overcome the DBP problem associated with chlorination. However, these alternative technologies do have their merits and drawbacks too.

In fact, scientists did not stop searching for innovative technologies for water disinfection. A wide range of innovative strategies have come into sight: copper-related methods, chitosan, TiO_2/UV, etc. After scrutiny of massive research papers, Xu Yang found that silver-related technologies may be the most promising solution and decided to choose it as his Ph.D. thesis topic.

This book is organized into five chapters. Chapter 1 provides a general overview of water disinfection, current practices, and their associated problems. This first chapter also explains why silver was chosen and how nanotechnology could be implemented to enhance the performance of silver.

Chapter 2 reviews the published materials on silver nanoparticles, from synthesis to antimicrobial activities, from function mechanism to laboratory-scale application.

More importantly, this chapter identifies the knowledge gaps and research questions Dr. Xu Yang would address in further chapters.

Chapters 3 and 4 describe the details of experimental setup and analysis of results. They cover four parts: (1) selection of nanosilver; (2) effects of water matrices on disinfection performance; (3) mechanism study of viral inactivation; and (4) study of continuous flow system integrated with nanosilver.

Final conclusion is summarized in Chap. 5, including recommendations for future study.

On behalf of other authors, I sincerely acknowledge the efforts of editorial team to make this book printed and published. We sincerely hope that you would find this book useful in your research/professional endeavors.

Singapore
May 2016

Prof. Jiangyong Hu

Parts of this thesis have been published in the following journal articles:

Yang Xu, Xiaona Chu, Jiangyong Hu and Say Leong Ong (2014). Disinfection performance of nanosilver and impacts of environmental conditions on viral inactivation by nanosilver. Water Science & Technology: Water Supply. 14(1):150–157.

List of Publications

Conference

Yang Xu, Xiaona Chu and Jiangyong Hu (2012). Performance evaluation of nanosilver disinfection against *E. coli* and MS2 coliphage. *World Water Congress & Exhibition*. Sept 16-21, 2012. Busan, Korea.

Yang Xu, Xiaona Chu and Jiangyong Hu (2012). Comparison of different types of nanosilver and application in continuous flow system. *21st KAIST-KU-NTU-NUS symposium on Environmental Engineering*. Jul 12–14, 2012. Kuala Lumpur, Malaysia.

Yang Xu, Xiaona Chu, Jiangyong Hu and Say Leong Ong (2011). Inactivation of *E. coli* And MS2 coliphage by silver nanoparticles combined With UV irradiation. *20th KAIST-KU-NTU-NUS symposium on Environmental Engineering*. Jun 26-28, 2011. Xitou, Taiwan.

Acknowledgments

First and foremost, I would like to express my most sincere gratitude to my supervisors, Prof. Say Leong Ong, and Associate Professor Jiangyong Hu. They have provided me with invaluable guidance and profound mentorship in the past few years, without which I could not complete this dissertation.

I am also grateful to the laboratory technicians in Water Science and Technology Laboratory. To Mdm Tan Xiaolan, Mr. S.G. Chandrasegaran, and Ms. Lee Leng Leng, I thank all their technical support and their efforts in laboratory maintenance.

Many thanks go to my colleagues in our research group. It is extremely fortunate for me to know all of you along this journey. I treasured all your encouragements when I was down and suggestions when I found no way out. All the joyful gatherings we had and the fun moments we shared decorated my entire Ph.D. and made it memorable.

Last but not least, my utmost gratitude goes to my wife and my parents. As the husband and son, I thank you all, for the care and understanding. Sometimes, I felt sorry that I was unable to be together with you on some of the big days. Now, I have finished it!

In fact, there are still many more people whom I need to thank but do not list here, including those names I do not even have the chance to know. I always feel warm and blissful from people around me, from little things that amazed me, from the shining sun and blowing breeze that soothed me. I love this world.

Contents

Abbreviations

AgNPs	Silver nanoparticles
ATCC	American Type Culture Collection
bio-AgNPs	Biologically synthesized silver nanoparticles
CFU	Colony-forming units
chem-AgNPs	Chemically synthesized silver nanoparticles
com-AgNPs	Commercial silver nanoparticles
ELISA	Enzyme-linked immunosorbent assay
ICP-OES	Inductively coupled plasma optical emission spectrometry
MBC	Minimum bactericidal concentration
MIC	Minimum inhibitory concentration
PFU	Plague-forming units
ROS	Reactive oxygen species
TEM	Transmission electron microscopy
TOC	Total organic carbon
USEPA	United States Environmental Protection Agency

List of Figures

List of Tables

Summary

Drinking water safety is always of great concern worldwide, and disinfection process of drinking water is vital to protect people against waterborne diseases caused by pathogens. Considering the intrinsic drawbacks of conventional technologies, there is an urgent need to develop a novel and robust alternative disinfection strategy that is capable of catering to increasing demand of safe drinking water.

In this study, silver nanoparticles (AgNPs or nanosilver) were explored systematically for the possibility of being a viable disinfectant. To achieve this goal, three kinds of AgNPs were first obtained from either market or laboratory synthesis. These three materials were characterized by transmission electron microscopy (TEM) and zeta potential analyzer, validating their nanoscale size. Subsequently, these AgNPs were compared in terms of antibacterial and antiviral activities against *Escherichia coli* and MS2 bacteriophage, respectively. Results indicated that biologically synthesized AgNPs (bio-AgNPs) should be selected based on the disinfection performance and ease of synthesis.

Secondly, effects of water matrices were studied in order to gauge the extent of bio-AgNPs in antiviral activity. Water matrices, including hardness (Ca^{2+} and Mg^{2+}), ionic strength, chloride, total organic carbon (TOC, as humic acid), and pH, were tested in this part of study. Results showed that disinfection capability of bio-AgNPs was highly sensitive to TOC, chloride, and pH. It suggested that concentrations of TOC and chloride might have to be controlled within suitable ranges for effective application of AgNPs in drinking water treatment. Certain pretreatment may be necessary.

Antiviral mechanisms of nanosilver were studied with MS2 bacteriophage as the virus model. Modes of action of nanosilver were investigated by verifying three mechanisms proposed by previous studies. Results demonstrated that direct contact of virus to nanosilver was the most important mechanism behind antiviral activities. Viral damage caused by nanosilver was also identified with regard to antigenicity damage and genome damage. It was found that antiviral performance was greatly associated with antigenicity damage, but less related to genome damage. This part of research implied that antiviral activities were mainly contributed by direct contact of virus to nanosilver, which may lead to severe damage to viral capsid (antigenicity).

A continuous flow system with nanosilver incorporated was finally designed and operated. This system was tested against both *E. coli* and MS2 bacteriophage in synthetic water. Dosage of nanosilver and strategies of backwash were studied for the purpose of improving the overall disinfection performance. It showed that backwash with air scour was able to regenerate the disinfection capacity by approximate 1-log removal in inactivation of *E. coli*. Increasing the dosage of nanosilver was another direct and effective strategy to improve the performance.

Keywords Disinfection · *Escherichia coli* · MS2 bacteriophage · Mechanism · Nanosilver

Chapter 1
Introduction

1.1 Drinking Water Safety

The significance of disinfection has been well recognized since one hundred years ago when epidemics (e.g. typhoid and cholera) were common and accounted for catastrophic death in the whole world. Drinking water treatment technology has been improving over time and a wide range of disinfectants and technologies have emerged and been applied by water utilities. Nevertheless, it was reported that 1.2 billion people did not have adequate safe drinking water, 2 billion had poor sanitation, and millions of people died annually from waterborne diseases directly (Montgomery and Elimelech 2007). It is known that many more are still suffering from water-related issues.

Pathogenic microorganisms transmitted through unsafe water are known to be a prime cause of these problems. Diarrhoeal disease, for instance, is believed to be mainly caused by a host of waterborne bacterial, viral and parasitic organisms. According to a report from World Health Organization (WHO 2004a), diarrhea alone accounts for an estimated 4.1 % of the total global burden of disease (measured in 'Disability-Adjusted Life Year') and is responsible for nearly 1.8 million deaths every year. Situations are even worse in developing countries where diarrhea is responsible for as high as 8.5 and 7.7 % of all deaths in Southeast Asia and Africa, respectively.

In developed countries, waterborne pathogen still remains as one of the main challenges and threats to public health. Despite advances in water management

I am also grateful for the support from IWA Publishing. Part of this thesis was reproduced from Y. Xu, X.-N. Chu, J.-Y. Hu and S.L. Ong (2014) "Disinfection Performance of Nanosilver and Impacts of Environmental Conditions on Viral Inactivation by Nanosilver". *Water Science & Technology* Vol. 14, pp. 150–157, with permission from the copyright holders, IWA publishing.

X. Yang, *A Study on Antimicrobial Effects of Nanosilver for Drinking Water Disinfection*, Springer Theses, DOI 10.1007/978-981-10-2902-8_1

practices, Centers for Disease Control and Prevention (CDC) in the United States reported 33 drinking water-associated outbreaks nationwide during 2009–2010, which comprised 1040 cases of illness, 85 hospitalizations and 9 deaths (CDC 2013). In Singapore, there were 127,402 attendances reported as acute diarrhoeal illnesses in 2012, an increase of 2.5 % over 2011 (MOH 2013).

1.2 Current and Emerging Waterborne Pathogens

There is a huge variety of infectious agents that are currently found to cause waterborne diseases, such as bacteria, viruses, fungi, protozoa, prions, helminthes, etc. (WHO 2003). Even worse is that more and more pathogens continue to emerge and threaten drinking water safety worldwide. According to the definition given by CDC, emerging pathogens are those that have increased the incidence of disease over the past two decades or threatened to increase in the near future (Ewald 1996). Recently several emerging waterborne pathogens have been identified, including *Legionella*, *Mycobacterium avium* complex, new enteric viruses (e.g. noroviruses), etc. (Szewzyk et al. 2000).

1.2.1 Bacterial Pathogens

According to WHO report, bacterial waterborne infections are one of the major cause of human morbidity and mortality around the globe (WHO 2003). In the US, species of enteric bacteria in the genera *Escherichia*, *Salmonella*, *Yersina* were identified as the main pathogens in waterborne disease outbreaks for the past two decades (Yoon and Hovde 2008). Among these, *Escherichia coli* O157:H7 was the most infamous bacterial pathogen due to its high risk of causing fatal infections in vulnerable persons, especially children and the elderly. The most recent outbreak of *E. coli* O157:H7 occurred in Ontario, Canada, in 2000, resulting in six deaths and over 2300 cases.

The bacterium *Legionella* is a typical example of emerging pathogens recently (Ngwenya et al. 2013). They are parasites of protozoa and widely distributed in ground water, wet soil (Riffard et al. 2001), and even in distribution and plumbing system biofilms (Robers and Keevil 1992). The latest statistics showed that from 2009 to 2010, there were 166 cases reported with 14 deaths caused by *Legionella* spp. in drinking water and other non-recreational water (Hilborn et al. 2013). On top of consumption of drinking water, human exposure to *Legionella* spp. is thought to be via inhalation of contaminated water aerosols, from cooling/air conditioning, for instance.

1.2.2 Viral Pathogens

Viruses are microorganisms that are composed of genetic material, deoxyribonucleic acid (DNA) or ribonucleic acid (RNA), along with a protective protein coat. It has been identified that many viruses can cause acute gastrointestinal disease, such as adenoviruses, noroviruses and rotaviruses.

A waterborne outbreak associated with norovirus was reported in a small village in Akita, Japan in 2005, where 29 people were infected by contaminated well water (Haramoto et al. 2012). Meanwhile, noroviruses as well as various viruses were frequently detected in surface water samples in Japan (Haramoto et al. 2008; Kitajima et al. 2010).

In fact, there are many more pathogens identified globally every year, such as protozoa and fungi. More essentially, the continued outbreaks of waterborne disease worldwide demonstrate that pathogenic organisms still pose a risk to drinking water hygiene and public health.

1.3 Selection of Disinfection Strategies

Disinfection process is a necessary step and an important practice to ensure water safety and protect public health. Scientists and engineers have been exploring various options to improve disinfection strategies in water treatment. Conventional disinfection strategies do have their advantages and are still employed in most water treatment systems. However, they do have some inevitable shortcomings that cannot be ignored in the long run.

1.3.1 Chlorine and Related Chemicals

Free chlorine has been proven to be a highly effective disinfectant and it has been used for decades around the globe because of its strong disinfection capability and low operation cost. Free chlorine is excellent at inactivating bacteria and viruses; and typically only 1–6 mg/L of chlorine is required in the final step of water treatment process (Crittenden et al. 2005). However, free chlorine was found to be ineffective in controlling some specific pathogens, such as *M. avium* (Taylor et al. 2000) and *Cryptosporidium parvum* (Carpenter et al. 1999). *C. parvum* is a protozoan parasite which can be found in feces-contaminated water. Once ingested, C. *parvum* would cause some symptoms including nausea and diarrhea in people with intact immune system. These symptoms can be fatal for some immunocompromised cases.

Another concern rising in recent years regarding chlorination is the potential generation of harmful disinfection by-products (DBPs). Free chlorine will inevitably react with organics present in water, producing halogenated chemical, such as trihalomethanes (THMs). Halogenated DBPs, such as THMs, are highly carcinogenic and strictly regulated by USEPA. Although free chlorine is still being used today around the world, scientists and engineers have been exploring measures to control the DBP generation and, at the same time, in developing novel non-halogen disinfection strategies.

Combined chlorine and chlorine dioxide (ClO_2) have been developed and applied as supplements which have been reported as better alternatives. These chlorine-containing chemicals indeed reduce formation of halogenated DBPs, such as THMs and haloacetic acids (HAAs). However, these strategies produce their unique DBPs instead. ClO_2 does produce two inorganic byproducts, chlorite (ClO_2^-) and chlorate (ClO_3^-), which are believed to have serious adverse health effects. For chloramine, the reaction of combined chlorine with organic matter commonly found in water may lead to the formation of N-Nitrosodimethylamine (NDMA), a byproduct classified as a "probable human carcinogen" by USEPA.

1.3.2 Ozone

Ozone is another commonly used chemical in drinking water disinfection. In terms of disinfection performance, ozone is even much more effective than free chlorine, chloramines and chlorine dioxide as ozone is more capable of inactivating viruses, *Cryptosporidium* and *Giardia.* As one of the most efficient chemical disinfectants, ozone requires a very short contact time and after decomposition, the only residual is dissolved oxygen. In addition, ozonation is capable of eliminating taste and odor compounds present in natural water body.

Despite its effectiveness for disinfection, ozone may generate intolerable DBPs such as bromate (BrO_3^-). Bromate has been considered a category 2B ("possibly carcinogenic to humans") carcinogen by the International Agency for Research on Cancer (IARC) and has also been banned by many countries worldwide. Apart from that, it is quite risky to handle ozone because it is highly corrosive and toxic. Therefore, generation of ozone should be carried out on site with high energy consumption and high level of operator skills. All in all the ozonation as disinfection is energy-intensive, capital-intensive and high level of maintenance is required (USEPA 1999).

1.3.3 Ultraviolet Irradiation

Ultraviolet (UV) radiation is an emerging disinfection strategy. Compared to free chlorine and ozone, UV possesses many attractive advantages: highly effective in controlling *C. parvum* oocysts; no harmful DBP has been identified so far; and lower requirements of operational cost, engineering expertise and infrastructure.

Mechanisms behind UV irradiation disinfection are believed to be DNA-related damage: Specifically UV induced thymine-thymine dimer generation in DNA, thereby interfering DNA reproduction and cell proliferation accordingly. This mechanism is direct and efficient; however post-treatment problems have been reported that such DNA damage could be photo-repaired and many bacteria and viruses previously inactivated by UV could reactivate and regrow (Zimmer and Slawson 2002). Repair mechanism involved has not yet been thoroughly known and is still being studied. Apart from the UV-induced DNA damage and repair, UV is also found to be ineffective in controlling certain notorious virus, adenovirus for instance. UV dosage of 40 mJ/cm^2 could only achieve approximately 1-log removal of adenovirus types 2, 5 and 41 (Baxter et al. 2007). On top of these, the efficacy of UV disinfection would be compromised when treating water containing UV-absorbing compounds. Presence of these components (such as humic acid) in the water decreases the ultraviolet transmittance (UVT) of water and thereby leads to lower UV dose delivery and less inactivation of target microorganisms (USEPA 2006). Moreover, UV would leave no disinfectant residue which is necessary to inhibit potential microorganism growth along the distribution systems.

More recently considerable efforts have been directed towards combined disinfection strategies, such as UV/Cl_2 and UV/O_3. Effective as they may seem to be, these combined systems are not thoroughly proven with respect to the negative aspects mentioned above.

While current disinfection strategies are effective in some aspects, they more or less have their own shortcomings due to their inherent limitations. With regulation becoming more and more stringent, there is an urgent need to develop an innovative disinfection strategy with robust and effective disinfection performance while avoiding or minimizing the negative aspects typically associated with the current state-of-the-art disinfection technologies.

1.4 Silver as Antimicrobial Agent

Antimicrobial effect of silver has been recognized and used for centuries. It is said that Alexander the Great (356–323 B.C.) drank only from silver vessel because of the antimicrobial effects of released Ag^+ ions and the Romans even wrote silver into the official Roman book of medicines (Silver et al. 2006). In 1965, Moyer and coworkers first suggested 5 % of silver nitrate be used to treat burns (Moyer et al. 1965). Following this, silver sulphadiazine was developed and widely applied in burn therapies, even in today's world (Storm-Versloot et al. 2010). As an antimicrobial agent, silver is attractive for many reasons: no unpleasant effects on the color, taste and odor of water; broad spectrum of activity against various microorganisms (Kim et al. 2007); difficulty for bacteria to develop resistance; and almost no adverse health impacts on human beings if low amount is digested. Silver and silver compounds are so attractive that there are numerous applications using it as antimicrobial agents: portable water filters; medical devices; clothing; and coatings for washing machines, refrigerators and food containers.

However, there are two main obstacles that prevent silver from being widely used for drinking-water disinfection. The first obstacle is the high price of silver. As a precious metal, silver is sold at 20.6 USD/oz. (price on 13/01/2014) currently in trade markets. The other obstacle is the uncompetitive disinfection performance. Compared to free chlorine or any other disinfectants used in water treatment plant, silver has not yet shown to provide fast removal of bacteria or virus so far.

To overcome these two shortcomings, one should investigate the way how silver is applied. Generally speaking, there are three main methods that silver species are employed as a biocide in current engineered systems: (1) silver-coated devices; (2) direct dosage of silver compounds; and (3) release of Ag^+ from silver electrode through electrolysis. The first method has limited Ag^+ released and low probability of pathogens reacting/contacting with silver or silver surface. The latter two methods may solve this problem by increasing probability of reaction with free form of silver; however they mainly depend on the antimicrobial effects of ionic form of silver, which is difficult to retain or recycle in current engineered system that inevitably leads to an increase in the capital cost.

Development of science in the past decade has clearly showed that electronic, optical and catalytic properties differ from the bulk metals when the particle size decreases to nanoscale (Niemeyer 2001; Moore 2006). Such novel properties made nanomaterials attractive and nanosilver is one of them. In fact, nanosilver has received attention not only in academic publication but also in manufacturing products for public use over last 100 years, even before the terminology 'nanomaterial' was invented (Nowack et al. 2011). Of the many nanotechnology-related consumer products, those which contain nanosilver were reported to be the largest (25 %) as well as the fastest growing category (Liu and Hurt 2010). These interests are mainly attributed to an intrinsic nature of nanosilver: large specific surface-to-volume ratio. This high specific surface area may enhance antimicrobial effects of silver over bulky silver or silver coating. In addition, nanosilver may be engineered into some kinds of materials that somehow can be retained within the system, such as magnetic $Fe_3O_4@Ag$ nanoparticles (Gong et al. 2007). In view of above, nanosilver may be the attractive solution to the challenges associated with disinfection practices mentioned above.

1.5 State-of-the-Art

Studies on disinfection performance of silver related materials have been progressing over the years. Some of these studies are significant and can be recognized as milestones in this field.

Silver ions were found to enhance the UV inactivation of MS2 coliphage by Butkus and his coworkers (Butkus et al. 2004, 2005). Log removals of MS2 coliphage were reported to be 0.25 and 1.75 by silver ion (0.1 mg/L, 130 min) and UV radiation (40 mJ/cm^2), respectively. Combination of these two would, however, achieve 3.25-log removal of MS2 coliphage, which indicated a synergy of 1.25-log removal. This finding is vital in enhancing the antiviral performance of UV radiation by adding only a small amount of aqueous silver.

After reporting this finding on bench scale, this research team further studied the feasibility of the silver/UV process for drinking water disinfection. There was a concern that some impurities commonly present in raw and treated drinking water may influence such silver/UV process. Results showed that parameters, such as turbidity (kaolin), hardness, alkalinity and pH, did not significantly reduce disinfection efficacy of the proposed silver/UV process. The other impurities such as chloride, phosphate and organic matters, on the other hand, were found to reduce the synergistic effect between silver and UV radiation disinfection. Nevertheless, silver/UV inactivation system was still beneficial for greater log removal (0.5-log difference) than UV alone. The authors also proposed the reasons for such adverse effects: (1) chloride and phosphate present in the system would react with aqueous silver that in turn reduced the effective concentration of silver ion and disinfection efficiency; (2) influence of some organic matters might lead to the increased UV_{254} absorbance, which resulted in a decrease in effective UV irradiation and thereby a corresponding decreases in the disinfection efficacy. Aside from evaluating the effects of water matrices, the authors also provided a feasible way of dosing silver ions in the proposed silver/UV system: Ag^+ could be continuously generated by electrolysis of silver electrode. An assessment of such operation and management costs involved was also discussed in their study and it suggested the proposed silver/UV could be economically beneficial.

Disinfection of *E. coli* and MS2 coliphage by aqueous silver combined with UVA has been explored by Kim and coworkers. They reported synergy between silver ion and irradiation of low energy, such as UVA or visible light instead of UV_{254}, for *E. coli* and MS2 coliphage inactivation (Kim et al. 2008). This finding made it possible to develop solar disinfection and other disinfection strategies which are safe, robust, economically beneficial and environmentally friendly. Mechanism behind the silver/UVA (or silver/visible light) disinfection has been proposed and examined. It is believed that silver ions and the thiol groups in cysteine likely to have undergone a photochemical reaction, forming silver-cysteine complex (R–S–Ag) and cysteine dimers (Eqs. 1.1–1.3) that could hinder the function of the affected protein. UVA and visible light played a vital role in accelerating this complexation reaction.

$$\mathrm{R{-}SH} + \mathrm{Ag}^+ \leftrightarrow \mathrm{R{-}S{-}Ag} + \mathrm{H}^+ \tag{1.1}$$

$$\mathrm{R{-}S{-}Ag} + h\nu \rightarrow \mathrm{R{-}S\cdot} + \mathrm{Ag} \tag{1.2}$$

$$\mathrm{R{-}S\cdot} + \mathrm{R{-}S\cdot} \rightarrow \mathrm{R{-}S{-}S{-}R} \tag{1.3}$$

On top of application of aqueous silver, silver nanoparticles (AgNPs) in disinfection, especially in virus inactivation, could also be interesting after scrutiny of recent studies. Biogenic silver nanoparticles (bio-AgNPs) were successfully immobilized in polyvinylidene fluoride (PVDF) membranes (De Gusseme et al. 2011).

As-prepared membrane was carefully examined in a continuous flow system where UZ1 bacteriophages were spiked into the feed water. This proposed bio-AgNPs/PVDF membrane achieved significant log removal (>3 log) of UZ1 bacteriophages with a water flux of 200 L/m^2. In addition, this membrane was capable of controlling the release of silver ion into the effluent so that the concentration of silver could be lower than the acceptable limit (0.1 mg/L) in the drinking-water (WHO 2004b).

Therefore AgNPs is a promising candidate as an effective disinfection strategy to be introduced into drinking-water treatment. However, there are four critical considerations that need to be addressed in order to realize this thought: (1) among different kinds of nanosilver, which nanosilver material is the best; (2) how feasible AgNPs could be in terms of disinfection when dealing with various and complex water matrices; (3) what is the mechanism of AgNPs disinfection; and (4) is there any possibility that such material could be incorporated into a continuous flow water treatment system. Only after these questions are adequately answered then we could apply silver for water treatment. A review of literature reveals that only limited studied focused on these fundamental questions have been carried out thus far. A lack of knowledge in this field motivates and inspires researchers to conduct further studies in order to realistically answer the four questions identified above.

1.6 Objectives and Research Scope

In view of the aforementioned opportunities and challenges in this field of AgNPs disinfection, this dissertation aims at systematically evaluating the feasibility of applying AgNPs in drinking water treatment process. The sub-objectives are: (1) to synthesize different kinds of silver nanoparticles and compare them in terms of disinfection performances; (2) to investigate the effects of water matrices on the disinfection; (3) to study the antiviral mechanism of AgNPs and (4) to apply AgNPs into a continuous flow system and improve overall performance by varying the operating conditions. Breakdown of these objectives is as follows:

Objective 1:

- To characterize three different kinds of silver nanomaterials synthesized, namely commercial silver nanoparticles (com-AgNPs), chemically-synthesized silver nanoparticles (chem-AgNPs) and biologically-synthesized silver nanoparticles (bio-AgNPs).
- To evaluate these three kinds of AgNPs with respect to inactivation of bacterial and viral models.

Objective 2:

- To investigate the effects of water matrices, including hardness (Ca^{2+} and Mg^{2+}), ionic strength, TOC, chloride ions and pH value.

Objective 3:

- To examine the modes of antiviral actions of AgNPs in terms of Ag^+ release, ROS generation and direct contact.
- To study antigenicity damage and genome damage after treatment of AgNPs.

Objective 4:

- To assess antibacterial and antiviral performance of continuous flow system containing AgNPs.
- To assess the effects of different backwash strategies and different dosage of AgNPs in continuous flow system.

With all these objectives completed, this study will provide the water industry with an in-depth understanding on the capacity, the extent and the mechanisms of nanosilver disinfection. This study will also be instructive in selecting the right nanosilver material and in optimizing nanosilver-involved disinfection system in the future.

The scope of this study is summarized in Fig. 1.1.

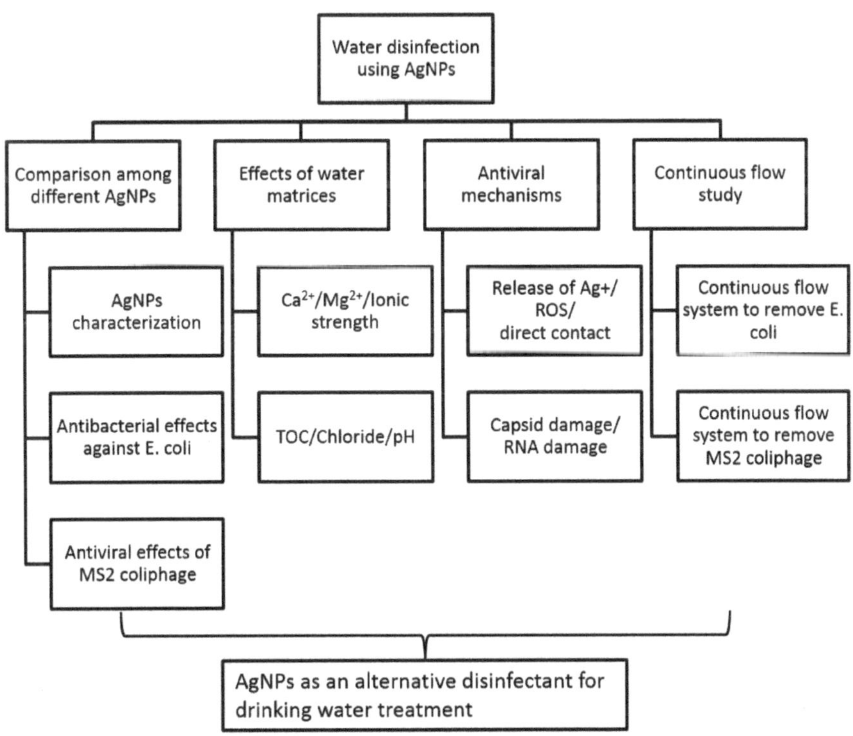

Fig. 1.1 Schematic framework of this study

1.7 Organization of Dissertation

This dissertation is organized into five chapters. Chapter 1 provides a general overview on current problems in drinking water disinfection and silver nanomaterials as a promising alternative of disinfectant.

Chapter 2 reviews the previous studies on AgNPs synthesis, antimicrobial effects of AgNPs as well as mechanisms behind such antimicrobial activities. Key contributions of past studies are highlighted in this chapter. More importantly, knowledge gaps and research needs are also identified.

Chapter 3 provides a detailed summary on materials, experimental design and the methodology adopted in this study.

Chapter 4 describes the results obtained in this study, covering four parts: namely (1) selection of nanosilver, (2) effects of water matrices, (3) mechanism study of viral inactivation by nanosilver and (4) study of continuous flow system with nanosilver. Analysis of the results and comparison with the past studies are also highlighted.

Finally conclusion from this study is summarized in Chap. 5. In addition, recommendations for future study are also provided.

References

Baxter CS, Hofmann R, Templeton MR, Brown M, Andrews RC (2007) Inactivation of adenovirus types 2, 5, and 41 in drinking water by UV light, free chlorine, and mlonochloramine. J Environ Eng-Asce 133:95–103

Butkus MA, Labare MP, Starke JA, Moon K, Talbot M (2004) Use of aqueous silver to enhance inactivation of coliphage MS-2 by UV disinfection. Appl Environ Microbiol 70:2848–2853

Butkus M, Talbot M, Labare M (2005) Feasibility of the silver-UV process for drinking water disinfection. Water Res 39:4925–4932

Carpenter C, Fayer R, Trout J, Beach MJ (1999) Chlorine disinfection of recreational water for *Cryptosporidium parvum*. Emerg Infect Dis 5:579–584

CDC (2013) Surveillance for waterborne disease outbreaks associated with drinking water and other nonrecreational water—United States, 2009–2010. Morb Mortal Wkly Rep (MMWR). Centers for Disease Control and Prevention

Crittenden JC, Trussell RR, Hand DW, Howe KJ, Tchobanoglous G (2005) Water treatment: principles and design, 2nd edn. Wiley

De Gusseme B, Hennebel T, Christiaens E, Saveyn H, Verbeken K, Fitts JP, Boon N, Verstraete W (2011) Virus disinfection in water by biogenic silver immobilized in polyvinylidene fluoride membranes. Water Res 45:1856–1864

Ewald PW (1996) Guarding against the most dangerous emerging pathogens: insights from evolutionary biology. Emerg Infect Dis 2:245–257

Gong P, Li H, He X, Wang K, Hu J, Tan W, Zhang S, Yang X (2007) Preparation and antibacterial activity of Fe_3O_4@Ag nanoparticles. Nanotechnology 18:285604

Haramoto E, Katayama H, Phanuwan C, Ohgaki S (2008) Quantitative detection of sapoviruses in wastewater and river water in Japan. Lett Appl Microbiol 46:408–413

Haramoto E, Kitajima M, Kishida N, Katayama H, Asami M, Akiba M (2012) Occurrence of viruses and protozoa in drinking water sources of Japan and their relationship to indicator microorganisms. Food Environ Virol 4:93–101

Hilborn ED, Wade TJ, Hicks L, Garrison L, Carpenter J, Adam E, Mull B, Yoder J, Roberts V, Gargano JW (2013) Surveillance for waterborne disease outbreaks associated with drinking water and other nonrecreational water—United States, 2009–2010. MMWR-Morb Mortal Wkly Rep 62:714–720

Kim J, Kuk E, Yu K, Park S, Lee H, Kim S, Park Y, Hwang C (2007) Antimicrobial effects of silver nanoparticles. Nanomed Nanotechnol Biol Med 3:95–101

Kim JY, Lee C, Cho M, Yoon J (2008) Enhanced inactivation of *E. coli* and MS-2 phage by silver ions combined with UV-A and visible light irradiation. Water Res 42:356–362

Kitajima M, Oka T, Haramoto E, Takeda N, Katayama K, Katayama H (2010) Seasonal distribution and genetic diversity of genogroups I, II, and IV noroviruses in the Tamagawa River, Japan. Environ Sci Technol 44:7116–7122

Liu JY, Hurt RH (2010) Ion release kinetics and particle persistence in aqueous nano-silver colloids. Environ Sci Technol 44:2169–2175

MOH (2013) Communicable diseases surveillance in Singapore 2012. Ministry of Health

Montgomery MA, Elimelech M (2007) Water and sanitation in developing countries: including health in the equation. Environ Sci Technol 41:17–24

Moore MN (2006) Do nanoparticles present ecotoxicological risks for the health of the aquatic environment? Environ Int 32:967–976

Moyer CA, Brentano L, Gravens DL, Margraf HW, Monafo WW (1965) Treatment of large human burns with 0.5 % silver nitrate solution. Arch Surg 90:812–867

Ngwenya N, Ncube EJ, Parsons J (2013) Recent advances in drinking water disinfection: successes and challenges. In: Whitacre DM (ed) Reviews of Environmental Contamination and Toxicology, vol 222. pp 111–170

Niemeyer CM (2001) Nanoparticles, proteins, and nucleic acids: biotechnology meets materials science. Angew Chem Int Edit 40:4128–4158

Nowack B, Krug HF, Height M (2011) 120 Years of nanosilver history: implications for policy makers. Environ Sci Technol 45:1177–1183

Riffard S, Douglass S, Brooks T, Springthorpe S, Filion LG, Sattar SA (2001) Occurrence of Legionella in groundwater: an ecological study. Water Sci Technol 43:99–102

Robers J, Keevil CW (1992) Immunogold and fluorescein immunolabelling of Legionella-pneumophila within an aquatic biofilm visualized by using episcopic differential interference contrast microscopy. Appl Environ Microbiol 58:2326–2330

Silver S, Phung LT, Silver G (2006) Silver as biocides in burn and wound dressings and bacterial resistance to silver compounds. J Ind Microbiol Biot 33:627–634

Storm-Versloot MN, Vos CG, Ubbink DT, Vermeulen H (2010) Topical silver for preventing wound infection. Cochrane Database Syst Rev

Szewzyk U, Szewzyk R, Manz W, Schleifer KH (2000) Microbiological safety of drinking water. Annu Rev Microbiol 54:81–127

Taylor RH, Falkinham JO, Norton CD, LeChevallier MW (2000) Chlorine, chloramine, chlorine dioxide, and ozone susceptibility of *Mycobacterium avium*. Appl Environ Microbiol 66: 1702–1705

USEPA (1999) Alternative disinfectants and oxidants guidance manual. EPA 815-R-99-014

USEPA (2006) Ultraviolet disinfection guidance manual for the final long term 2 enhanced surface water treatment rule. U.S. Environmental Protection Agency

WHO (2003) Quantifying selected major risks to health. The World Health Report 2002. World Health Organization, Geneva

WHO (2004a) Burden of disease and cost-effectiveness estimates. World Health Organization
WHO (2004b) Guidelines for drinking water quality. World Health Organization
Yoon JW, Hovde CJ (2008) All blood, no stool: enterohemorrhagic *Escherichia coli* O157: H7 infection. J Vet Sci 9:219–231
Zimmer JL, Slawson RM (2002) Potential repair of *Escherichia coli* DNA following exposure to UV radiation from both medium- and low-pressure UV sources used in drinking water treatment. Appl Environ Microbiol 68:3293–3299

Chapter 2
Literature Review

Silver nanoparticles (AgNPs or nanosilver) material has been receiving increasing attention in the recent decade. For example, AgNPs and its application has been studied and exploited to a large extent. In fact, it was reported that nanosilver had a history of approximately 120 years of usage, mainly under the term 'colloidal silver' (Nowack et al. 2011). In this proposed study, previous literature was critically reviewed in order to obtain a comprehensive and systematic perspective of this field. These following sections will cover different synthesis of silver nanoparticles and relevant AgNPs-contained materials, antimicrobial effects thereof and mechanisms involved in antimicrobial effects of AgNPs.

2.1 Synthesis of Silver Nanoparticles

Nanoparticles are particles which range from 1 to 100 nm in size. Numerous studies conducted in the past years demonstrated that these materials behave entirely different from their bulky counterparts with regard to the optical, electronic and catalytic properties, etc (Niemeyer 2001; Sun and Xia 2002; Moore 2006). These properties were believed to strongly correlate to the shape, size and substructure of metal nanoparticles. It was stated that properties of nanoparticles could be finely tuned by controlling these parameters (Sun and Xia 2002; Xia et al. 2008). Therefore, it is necessary to summarize synthesis routes and achievements in AgNPs synthesis. In general, there are three key considerations in the synthesis of AgNPs. They are: (1) morphology of AgNPs (mainly size and shape); (2) stability of AgNPs; and (3) feasibility and cost of large-scale production of AgNPs. The following sections would be focusing on these aspects.

X. Yang, *A Study on Antimicrobial Effects of Nanosilver for Drinking Water Disinfection*, Springer Theses, DOI 10.1007/978-981-10-2902-8_2

2.1.1 Chemically-Synthesized Silver Nanoparticles

Sun and Xia managed to synthesize silver nanocubes of tunable size, 50–120 nm, with PVP as capping agents (Sun and Xia 2002). The size of such AgNPs could be determined by simply adjusting the molar ratio of capping agents. The primary reaction involved a so-called polyol process (reduction of silver nitrate with ethylene glycol at 160 °C) in which ethylene glycol served as both reductant and solvent. This publication served as a basis for controlling shapes, sizes and structures of AgNPs and disclosed the wide possibilities of wet chemistry methods that could be used in this area.

Another shape of silver nanoparticles, nanoprism, was successfully synthesized by Mirkin's group (Jin et al. 2001; Metraux and Mirkin 2005; Millstone et al. 2009). The first reliable synthesis of silver nanoprism was achieved through the photo-induced conversion from silver nanoparticles (Jin et al. 2001). It was reported that prismatic structure of AgNPs could be obtained by irradiation of a conventional 40-W fluorescent light on previously prepared spherical particles (8.0 $\pm$ 1.7 nm). As-prepared silver nanoprism had a size with 100 nm in length and 16 nm in thickness. Furthermore, a variety of radiation wavelengths had been found to be capable to induce this conversion, from visible lights to UV irradiation. These achievements were one of the first few synthesis methods that combined light irradiation and chemical reduction, which even paved the way to a broader area of AgNPs synthesis.

Compared with photo-induced synthesis, thermal synthesis (or chemical reduction methods) of silver nanoprism were reported to be able to achieve AgNPs of similar shape and size (Metraux and Mirkin 2005). However, information on purification and stability of such silver nanoprism was not discussed.

Truncated triangular silver nanoplates were slightly different from nanoprism; and synthesis of this specific structure was also appealing to consider due to a proposed mechanism in which different crystal facets may induce different extent of antimicrobial effects (Pal et al. 2007). Typical procedures for synthesis and characterization of truncated triangular silver nanoplates were reported by Chen and Carroll (2002). Three phases were involved in the synthesis route: seeding, growth and aging. First, sodium borohydride ($NaBH_4$) was used to reduce Ag^+ into Ag(0) with sodium citrate as stabilizing agent. After this seeding phase, more Ag^+ was reduced by more reducing agents and grew onto these AgNP seeds with cetyltrimethlyammonium bromide (CTAB) as capping agent. In the last step, the silver nanoparticles were required to undergo aging in order to allow the particles grow to the expected structure, truncated triangular nanoplates. Centrifugation was required to further remove undesired structures (e.g. rod, sphere, cube, etc.) so as to purify the synthesized nanomaterials.

Among all the nanostructures of silver nanoparticles, studies on spherical particle probably have gained the most attention due to ease of synthesis, purification and application. Great contribution was made by Sondi and his team in developing an easy method to prepare highly concentrated and stable silver nanoparticles

(Sondi et al. 2003). With Daxad 19 as stabilizing agent and ascorbic acid as reducing agent, the as-prepared silver nanoparticles has excellent aqueous stability at a wide range of pH values ranging from 2 to 10. Another advantage of this method is the capability of tuning the particle size between 10 and 30 nm based on the reaction time. This study is vital because high concentration of silver nanoparticles with tunable size is a prerequisite for the comprehensive study of antimicrobial effects of silver nanoparticles (Fig. 2.1).

Most recent advances in AgNP synthesis focus not only on morphology of AgNPs, but on more diverse functionalities as well. Silver was coupled with Fe_3O_4 to accomplish electromagnetic Ag-Fe_3O_4 nanoparticle, which facilitated concentrating and recycling of precious silver (Gong et al. 2007; Li et al. 2014). Silver was also doped into nano-titanium oxide and carbon nanotube to enhance their

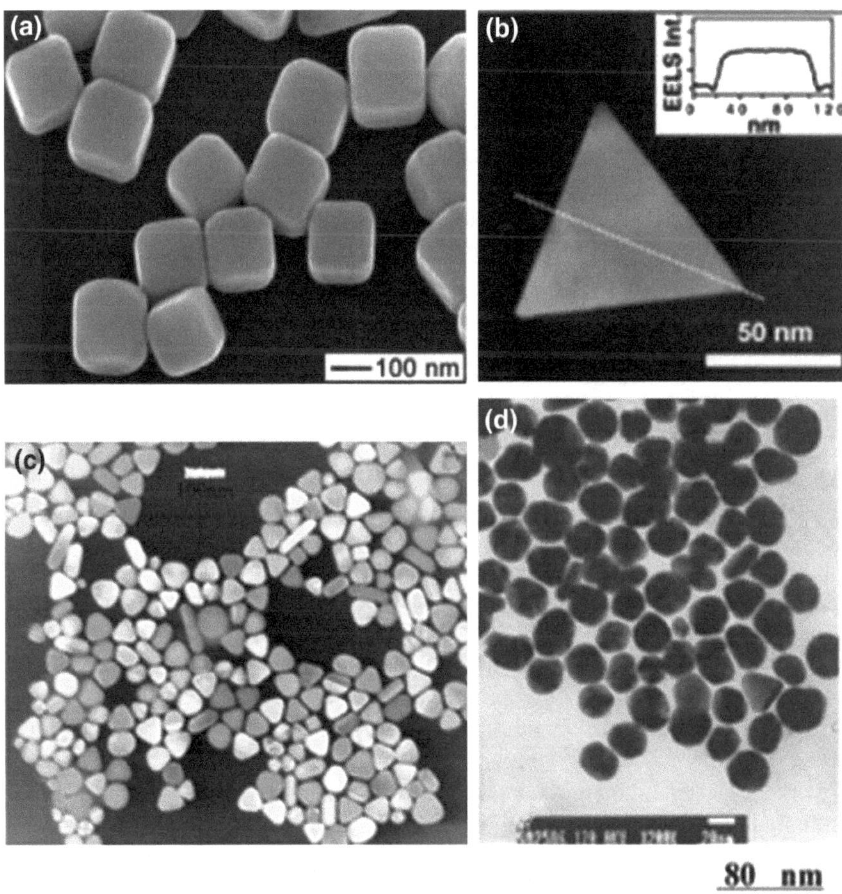

Fig. 2.1 TEM images of different morphologies of silver nanoparticles. **a** nanotubes (Sun and Xia 2002); **b** nanoprisms (Jin et al. 2001); **c** truncated triangle (Chen and Carroll 2002); and **d** stable and adjustable silver nanoparticles (Sondi et al. 2003)

photocatalytic efficiency, showing potential of application in water decontamination (Kim and Song 2014; Ziemkowska et al. 2014). Studies relevant to water disinfection are reviewed in the next section.

Shapes and sizes of silver nanoparticles can be finely tuned; monodispersity can be well achieved. All these advantages rely on the flexibility and diversity of chemicals and synthesis methods. However, the drawbacks of these methods are undeniable: aqueous stability of particles in the long run and involvement of special technology, such as light irradiation. Additionally, intensive chemical usage is a further unfavorable factor, posing some adverse impacts on vulnerable ecological environment.

2.1.2 Biologically-Synthesized Silver Nanoparticles

Apart from chemical synthesis, synthesis using biological agents is another vital alternative to produce nanosilver with unique properties. Compared to chemical synthesis, bio-synthesis makes use of biological agents to synthesize and stabilize AgNPs, avoiding usage of toxic chemicals. Synthesis of nanosilver using chemicals benefits from flexibility and diverse method options in laboratory; scaling up such synthesis, however, may be faced with inevitable problems, including high capital cost; high maintenance cost and potential toxicity to eco-system. Biological synthesis methods may make a difference in overcoming these shortcomings typically associated with chemical preparation methods.

Sintubin and co-workers reported that lactic acid bacteria, *Lactobacillus* spp. were capable of reducing Ag^+ into silver nanoparticles (Sintubin et al. 2009). Gram-positive bacteria including *Lactobacillus* spp. and *Enterococcus faecium*, and gram-negative bacteria such as *Pseudomonas aeruginosa* and *Escherichia coli* have been comprehensively screened for nanosilver production. As shown in comparison, *Lactobacillus* spp. was outstanding for its competitive capability in generating silver nanoparticles of smaller size (10–20 nm), more stable dispersion and higher yield. As-prepared nanosilver was situated widely in the cell, in the outer layer of cell and on the cell wall, which is believed to be an advantage in avoiding particle aggregation.

Silver nanoparticles of 10–50 nm could be synthesized by reacting Ag^+ ions with biomass of *Brevibacterium casei* (Kalishwaralal et al. 2010). Such synthesis route required less time (24 h only) and only benign conditions (37 °C). On top of successfully identifying the presence of nanoparticles, Kalishwaralal and his co-workers also found that protein or functional groups may play an important role and initialize this reaction based on Fourier Transform Infrared Spectroscopy (FTIR) results. This study disclosed the secret of bio-AgNPs and may lead to area that could actively select the best microorganism of interest.

Production of highly concentrated nanosilver materials was reported by Juibari et al. (2011). Juibari's team isolated a native extremophilic *Ureibacillus thermosphaericus* strain from hot springs. Interestingly, such bacteria strain was able to produce AgNPs of adjustable size (10–80 nm) by adjusting the initial concentration of Ag^+ ions as well as the temperature (60–80 °C). This study is important in realizing one idea that non-fastidious and native bacteria could be used directly to synthesize AgNPs at high concentration ($C_{Ag,\ total} = 0.1$ M). And impressively, the size of AgNPs (13–75 nm) could be easily tuned by adjusting the concentration of $AgNO_3$ and reaction temperature. However, authors did not monitor the long-term stability of such AgNPs in water.

Ahmad and co-workers successfully achieved extracellular biosynthesis of silver nanoparticles using the fungus *Fusarium oxysporum* (Ahmad et al. 2003). With direct addition of $AgNO_3$ into a solution of *F. oxysoporum* biomass, silver nanoparticles could be obtained after approximately 72 h. TEM images showed that the silver nanoparticles were in a size range of 5–50 nm with severe aggregation. In spite of problem of aggregation, Ahmad's study was one of the milestones that expanded to fungi with regard to bio synthesis of AgNPs. After that, similar results had been reported using two other types of fungi, *Aspergillus fumigatus* (Bfilainsa and D'Souza 2006) and *Fusarium semitectum* (Basavaraja et al. 2008) (Fig. 2.2).

Not only limited to bacteria and fungi, biosynthesis of silver nanoparticles has been expanded to utilizing biomass from yeast (Kowshik et al. 2003), algae

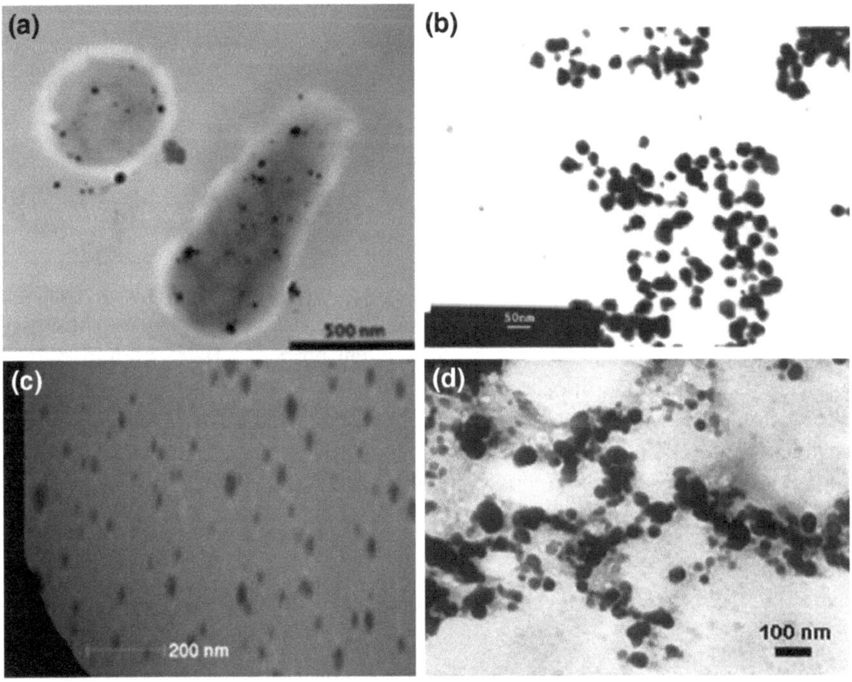

Fig. 2.2 TEM images of AgNPs synthesized from **a** *Lactobacillus* spp. (Sintubin et al. 2009); **b** *Brevibacterium casei* (Kalishwaralal et al. 2010); **c** *Ureibacillus thermosphaericus* strain (Juibari et al. 2011); and **d** *Fusarium oxysporum* (Ahmad et al. 2003)

(Govindaraju et al. 2008) and plants (Rajasekharreddy et al. 2010; Lukman et al. 2011; Veerasamy et al. 2011) as well. Biogenic silver nanoparticles, compared to chemically synthesized, is still new but is developing rapidly in the last five years. Despite the failure of controlling shape, size and purity of AgNPs, biological synthesis is an interesting alternative for chemical methods due to the ease of operation and green synthesis. Additionally the presence of biomass or biological agent may be better to stabilize AgNPs (Sintubin et al. 2009) with better bio-compatibility and less toxicity than heavily used industrial chemicals. It is believed that by choosing the proper biological system, biogenic AgNPs are able to achieve quality close to its peer and find the balance between green synthesis and functionalization (Sintubin et al. 2012).

It is evident from literature that both chem- and bio-synthesis, are intentionally limited to a term 'wet chemistry method'. Being developed for over the past decade, synthesis of AgNPs has also been extended to microwave-aided synthesis, UV-photolysis, electrodeposition, and dry photochemical growth, etc. Nevertheless, it is still worth mentioning that wet chemistry method is by far the most mature and commonly used technique for fabricating AgNPs, considering the capital cost, scalability and feasibility of application (Majdalawieh et al. 2014).

Comparison of chem- and bio-synthesis poses two distinct options: chemical synthesis benefits from finely control of shape/size and monodispersity while biological synthesis benefits from green process and that biomass may serve as a better stabilizing agent. This raises a concern on choosing the right route. Typical chem- and bio-synthesis routes should be selected and compared in view of water disinfection.

2.2 Antimicrobial Effects of Silver Nanoparticles

Silver products have been used for various antimicrobial applications for over thousands of years (Silver et al. 2006); and it won't be surprising to know that nanosilver materials naturally share such excellent property. There are hundreds of publications that involve the antimicrobial effects of AgNPs over the past decade. Specific topics relevant to water disinfection are reviewed below, including effects of silver nanoparticles' properties and water matrices on antimicrobial activities, along with practices of continuous system with AgNPs.

2.2.1 Effects of Silver Nanoparticles' Properties

Morones and co-workers first studied the size effects of nanosilver on antibacterial effects (Morones et al. 2005). Not only characterizing initial AgNPs, those researchers also carefully focused on characterization of the AgNPs after the experiments, which were found attached on the bacterial cell membrane. Two interesting findings were reported: (1) crystal facet {111} of AgNPs was more likely to attach to the cell membrane, possibly due to the higher atom density and higher reactivity compared to other facets; and (2) most of nanosilver interacting with *E. coli* was in the size range of 1–10 nm. These two findings were consistent: silver {111} facet was higher at the density of atoms while 1-10 nm particles had higher percentage of surface atoms than larger particles. Both findings illustrated that more atoms on the surface was favorable to reactivity of AgNPs, which makes tuning of nanosilver morphology a practical method to enhance the disinfection efficiency.

Inspired by their peers, Pal and coworkers further examined the effects of various shapes of AgNPs on its antimicrobial activities (Pal et al. 2007). By comparing three different kinds of AgNPs, namely spheres, rods and truncated triangle, the researchers concluded that antimicrobial effect of silver nanoparticles was dependent on their shape and nanosilver with truncated triangular shape had the best performance over spherical and rod-shaped counterparts. Proposed reason was in line with Morones et al. (2005): the top basal plane of a truncated triangular nanoplate has a {111} surface, which had higher reactivity to the bacterial surface.

Capping agents, generally, will determine the surface charge of AgNPs that in turn has an impact on antimicrobial ability of AgNPs as described by El Badawy et al. (2011). These researchers studied four kinds of AgNPs with different capping agents, namely H_2-AgNPs, citrate-AgNPs, PVP-AgNPs and BPEI-AgNPs. These four AgNPs were delicately synthesized so that they had similar size (10–18 nm) yet with distinct surface charge [from the lowest (−40 mV) of citrate-AgNPs to the highest (+40 mV) of BPEI-AgNPs]. Additionally the authors purified these four AgNPs continuously to chip away the impact of released Ag^+. With such careful experiment design, the effect of surface charge could be realistically studied. They reported that the more positive the surface charge was, the higher toxicity it showed. It was explained by electrostatic barrier between AgNPs and the bacteria used in their study. As we may know, surface of bacteria are negatively-charged because of dissociation of organic groups on the cell membranes, such as Singapore, carboxyl, phosphate and amino groups. Thus, electrostatic barrier was reduced with decrease of negative zeta potential of AgNPs and positively-charged AgNPs even turned such repulsion into attraction, thereby allowing higher chance of interactions. Such physical interactions (i.e. direct contact) were therefore believed to be the primary mechanism for AgNPs toxicity. That is, surface charge was another factor that would affect antimicrobial performances of nanosilver.

2.2.2 Effects of Water Matrices

Diversity of water matrices is always a key consideration when discussing drinking water treatment, especially for water disinfection. Knowing which parameter may enhance or hinder the antimicrobial effects of nanosilver would equip us with better knowledge to apply nanosilver for water treatment. There are a few publications that have already focused on effects of water matrices on antimicrobial activity of AgNPs. Fabrega et al. (2009) showed that the presence of humic substances was able to completely remove the bactericidal effects of nanosilver. Two milligrams per litre of AgNPs would lead to approximate 80 % inhibition of *Pseudomonas fluorescens*; by contrast, however, no inhibition occurred under same conditions except that 10 mg/L of humic acids was added to the system. Fabrega and coworkers proposed that humic substances might form a film on the NP, thereby reducing the chance of interaction between AgNPs and microbes. This proposed reason was in line with the understanding derived from previous studies (Sondi and Salopek-Sondi 2004; Shrivastava et al. 2007) and those reported in other nanoparticle-related investigation (Baalousha et al. 2008; Diegoli et al. 2008). On the other hand, humic substances could also act as an antioxidant by reacting with any ROS, which is another frequently cited mechanism of AgNPs toxicity (Fabrega et al. 2009). Therefore humic substances could worsen the antibacterial performance regardless of which mechanism is primary.

Jin and coworkers' study was also focusing on such "surface-related mechanism" (Jin et al. 2010). Through study of various combinations of ubiquitous cations and anions on bacterial inactivation of AgNPs, Jin's team found that presence of Ca^{2+}/Mg^{2+} ions slightly enhanced the antibacterial performance of nanosilver. Such observation was attributed to these divalent cations being able to form "ion bridge" between negatively-charged AgNPs and lipopolysaccharide (LPS) on cell membrane. The latter is also negatively charged for gram-negative bacteria, *Pseudomonas putida*, used in their study. It was thought that such electrostatic interaction might change the cell membrane permeability and facilitate the transfer of silver nanoparticles into the cells that enabled more AgNPs access to sites of action. This study clearly illustrated that type of electrolytes in water had a direct impact on nanosilver disinfection and study on this area would be a must for further application of AgNPs in water treatment.

On top of Ca^{2+} and Mg^{2+}, other electrolytes also have some impacts on antimicrobial effects of AgNPs. Chloride (Cl^-) is ubiquitous in drinking water and well known for its fast precipitation with Ag^+. Back to 1998, Gupta and coworkers studied that effects of chloride on toxicity of Ag^+ against *E. coli* (Gupta et al. 1998). It was found that when increasing concentration of Cl^- up to 30 g/L, toxicity of Ag^+ could be enhanced, even when inactivating Ag-resistant species, *E. coli* J53 (pMG101). Such enhancement was believed to be caused by formation of soluble complex ions, $AgCl_2^-$ and $AgCl_3^{2-}$, at high concentration of Cl^-. This finding was further validated by Choi et al. (2008) and Levard et al. (2013), in which the AgCl colloids (or soluble $AgCl_x^{(x-1)-}$ species) were found to have as significant toxicity as Ag^+ and AgNPs (Choi et al. 2008; Levard et al. 2013). Most of the concentrations of Cl^- used in the above studies were more than 10 g/L, which was applicable to wastewater treatment and extreme eco-environments, but was rare in drinking water treatment.

Effect of pH is another factor worth considering in the area of water disinfection. Studying the growth of planktonic *P. fluorescens* with AgNPs at different pH (6, 7.5 and 9), Fabrega and her team found that alkaline condition (pH = 9) was preferred in the hope of enhanced toxicity (Fabrega et al. 2009): 90 % decrease of growth in pH 9 while only 10 % decrease in pH 6 at the same dosage of 2 ppm AgNPs. Interestingly, varying pH from 6 to 9 did not change the particle size distribution significantly ($P > 0.05$). Therefore, there is still a gap explaining such pH-related toxicity. Furthermore, El Badawy and coworkers conducted a more comprehensive study on the effect of pH on the particle size of AgNPs (El Badawy et al. 2010). It was reported that whether pH would affect the size was dependent on which capping agent was used. Typically size of sterically stabilized AgNPs (e.g. PVP-coated) tends to remain while sizes of other AgNPs (e.g. citrate-coated) were affected by pH change.

2.2.3 *Antibacterial and Antiviral Effects of Silver Nanoparticles*

Since Sondi and Salopek-Sondi first studied silver nanoparticles for antibacterial effects on *E. coli* in 2004, there has been a boost of publications that focus on the antimicrobial effects against different microorganisms. Some of the typical and important studies are summarized and discussed below.

Sondi and Salopek-Sondi first studied silver nanoparticles as antimicrobial agent with *E. coli* as a model for Gram-negative bacteria (Sondi and Salopek-Sondi 2004). Silver nanoparticles used in this study was synthesized through reduction of silver nitrate by ascorbic acid (Sondi et al. 2003); and the inhibition experiments were done both in plates and liquid LB medium. Seventy percent of inactivation was reported when 10 mg/L of nanosilver was added whereas nearly 100 % inactivation was observed when 50 mg/L was dosed. Growth inhibition of *E. coli* was successfully observed in experiments of dosing nanosilver into liquid LB medium as well. Nanosilver spread both in and outside of the cell observed by SEM and further TEM images showed that membrane of bacteria was pitted, leaking out the plasma. Equipped with their previous knowledge (Stoimenov et al. 2002), the authors speculated the antimicrobial effects were induced by contact of nanosilver with membrane, increasing permeability and causing cell death thereafter. Although antimicrobial capacity of AgNPs was well reported in this study, its full potential was not well exploited yet.

Other than mechanisms, studies also focused on the synthesis of novel material. Gong and co-workers successfully synthesized magnetic Fe_3O_4@Ag nanoparticles (Gong et al. 2007). Such particles were of sizes ranging from 40 to 80 nm and proven to be able to inhibit the growth of *E. coli* (gram-negative), *Staphylococcus epidermidis* (gram-positive) and *Bacillus subtilis* (spore bacteria). These magnetic

Table 2.1 Summary of literature on the antimicrobial and antiviral effects of AgNPs

Micro-organism	Capping agents	Conc.[a] (mg/L)	Contact time	Major effects	Highlights	References
Bacteria						
E. coli	Citrate/CTAB	0.01–1	0–26 h	Inhibition[b]	Sphere/rod/triangular AgNPs. Triangular was the best	Pal et al. (2007)
E. coli	ND[c]	25–100	ND[c]	Inhibition[b]	AgNPs <10 nm attached to cell membrane. {111} was significant	Morones et al. (2005)
E. coli	Biomass	12.5	2 h	7-log removal	Bio-AgNPs. Antibacterial capacity of AgNPs was similar to Ag^+	Sintubin et al. (2011)
E. coli	Triton X-100	10–100	ND[c]	MIC: 70 mg/L	Magnetic Fe_3O_4@Ag. It could be recycled and reused	Gong et al. (2007)
E. coli	Un-coated	0.2–33 nM	24 h	MIC > 3.3 nM	Bio-AgNPs	Kim et al. (2007)
E. coli	ND[c]	200 as Ag	1 h	6 log removal	Ag@Magnetic hybrid colloids	Park et al. (2013)
E. coli	Saccharide	ND[c]	24 h	MIC > 3.38 mg/L	Ag^+ was better than AgNPs in antimicrobial performance	Panacek et al. (2006)
Klebsiella pneumoniae	Saccharide	ND[c]	24 h	MIC > 6.75 mg/L	Ag^+ was better than AgNPs in antimicrobial performance	Panacek et al. (2006)
Pseudomonas aeruginosa	ND[c]	25–100	ND[c]	Inhibition[b]	AgNPs <10 nm attached to cell membrane. {111} was reported to be significant	Morones et al. (2005)
Pseudomonas aeruginosa	Biomass	1–500	ND[c]	MBC: 200 mg/L	Bio-AgNPs. Antibacterial capacity of AgNPs was similar to Ag^+	Sintubin et al. (2011)
Pseudomonas aeruginosa	Saccharide	ND[c]	24 h	MIC > 3.38 mg/L	Ag^+ was better than AgNPs in antimicrobial performance	Panacek et al. (2006)
Pseudomonas putida	ND[c]	0–50	24 h	IC_{50} >50	Ag^+ was better than AgNPs in antimicrobial performance	Jin et al. (2010)

(continued)

Table 2.1 (continued)

Micro-organism	Capping agents	Conc.[a] (mg/L)	Contact time	Major effects	Highlights	References
S. typhus	ND[c]	25–100	ND[c]	Inhibition[b]	AgNPs <10 nm attached to cell membrane. {111} was reported to be significant	Morones et al. (2005)
Vibrio cholerae	ND[c]	25–100	ND[c]	Inhibition[b]	AgNPs <10 nm attached to cell membrane. {111} was reported to be significant	Morones et al. 2005)
Bacillus spp	Citrate/PVP/BPEI	0.002–1	5 days	Inhibition[b]	BPEI-coated was the best because of positive charge	El Badawy et al. (2011)
Bacillus subtilis	ND[c]	0–50	24 h	IC50 < 50	Ag^+ was better than AgNPs in antimicrobial performance	Jin et al. (2010)
Bacillus subtilis	Triton X-100	10–100	ND[c]	MIC: 70 mg/L	Magnetic Fe_3O_4@Ag. It could be recycled and reused	Gong et al. (2007)
Enterococcus faecalis	Saccharide	ND[c]	24 h	MIC > 13.5 mg/L	Ag^+ was better than AgNPs in antimicrobial performance	Panacek et al. (2006)
Enterococcus faecium	Saccharide	ND[c]	24 h	MIC > 13.5 mg/L	Ag^+ was better than AgNPs in antimicrobial performance	Panacek et al. (2006)
Staphylococcus aureus	Biomass	1–500	ND[c]	MBC: 100 mg/L	Bio-AgNPs. Antibacterial capacity of AgNPs was similar to Ag^+	Sintubin et al. (2011)
Staphylococcus aureus	Un-coated	0.2–33 nM	24 h	MIC > 33 nM	Bio-AgNPs	Kim et al. (2007)
Staphylococcus aureus	Saccharide	ND[c]	24 h	MIC > 6.75 mg/L	Ag^+ was better than AgNPs in antimicrobial performance	Panacek et al. (2006)
Staphylococcus epidermidis	Triton X-100	10–100	ND[c]	MIC: 60 mg/L	Magnetic Fe_3O_4@Ag. It could be recycled and reused	Gong et al. (2007)
Staphylococcus epidermidis	Saccharide	ND[c]	24 h	MIC > 1.69 mg/L	Ag^+ was better than AgNPs in antimicrobial performance	Panacek et al. (2006)

(continued)

Table 2.1 (continued)

Micro-organism	Capping agents	Conc.[a] (mg/L)	Contact time	Major effects	Highlights	References
Viruses						
Adenovirus 3	Tannic acid	50	2 h	>70 % inhibition	AD3 DNA damage observed	Chen et al. (2013)
H1N1 influenza A virus	Glucose and Chitosan	400 as Ag	1 h	Nearly 100 %	AgNPs/Chitosan	Mori et al. (2013)
H3N2 influenza virus	Tannic acid	50	2 h	2.5 log removal	Inactivation in vivo was also reported on mice	Xiang et al. (2013)
Herpes simplex virus type 1	Fungi biomass	10	24 h	80 % inhibition	Myco-AgNPs	Gaikwad et al. (2013)
Herpes simplex virus type 2	Fungi biomass	10	24 h	50 % inhibition	Myco-AgNPs	Gaikwad et al. (2013)
Human parainfluenza virus type 3	Fungi biomass	10	72 h	80 % inhibition	Myco-AgNPs	Gaikwad et al. (2013)
MS2 coliphage	ND[c]	200 as Ag	1 h	2 log removal	Ag@Magnetic hybrid colloids. Recyclable	Park et al. (2013)
UZ1 bacteriophage	Biomass	5.4	1 min	4.2 log removal	Bio-AgNPs	De Gusseme et al. (2009)

[a]The unit of concentration is mg/L unless specified
[b]Inhibition characterized by unquantified growth delay curve
[c]ND: Not described

Fe_3O_4@AgNPs are attractive because it could be recycled by magnet after interacting with bacteria and reused, providing a new route to compensate for the high price of silver.

Diversity of antimicrobial effects of AgNPs was also extended to viruses. For example, synthesis of AgNPs using fungi biomass was developed and its antiviral effects was evaluated by Gaikwad and co-workers (Gaikwad et al. 2013). Size of silver nanoparticles obtained was 20–40 nm depending on which production systems (i.e. fungi strains) were used. Silver-virus interaction was found to be size-dependent: smaller particles tended to improve the inhibition effects. This finding was in line with previous antibacterial study (Morones et al. 2005), again showing the significance of AgNP size. With contact time (24–72 h), approximate 50–80 % of herpes simplex virus (both type 1 and 2) was inhibited by 10 mg/L of AgNPs. This study did make contributions to the extension of antimicrobial spectrum of nanosilver; however, long reaction time and relatively poor inhibition performance limited application in drinking water treatment process. Increasing the dosage of nanosilver might be a solution in practical use.

There is abundant literature on the antibacterial and antiviral effects of nanosilver in the past decade and relevant studies are summarized in Table 2.1. The majority of literature reviewed here merely focus on inhibition effects with long reaction period and high dosage of AgNPs without considering the capital cost and time requirement needed for water treatment. This gap serves as a purpose of investigating the antibacterial and antiviral effects of AgNPs in terms of both efficiency and cost; that is, shorter contact time, higher antimicrobial performance and relatively lower dosage of AgNPs.

2.3 Antimicrobial Mechanism of Silver Ions and Silver Nanoparticles

Understanding mechanisms of how nanosilver inactivates bacteria and viruses is a crucial route to enhance its disinfection performance. In this section, mechanisms of Ag^+ disinfection are first discussed since nanosilver is always associated with silver ions released from the surface. Literature on intrinsic antimicrobial nature of nanosilver is covered subsequently.

2.3.1 Mechanisms of Aqueous Silver Disinfection

Bactericidal effect of silver ion has been studied for decades (Russell and Hugo 1994); and interaction with thiol groups of the L-cysteine is the most widely known mechanism (Liau et al. 1997). Silver ion has an extremely strong affinity with sulfur ion (S^{2-}) (Eq. 2.1).

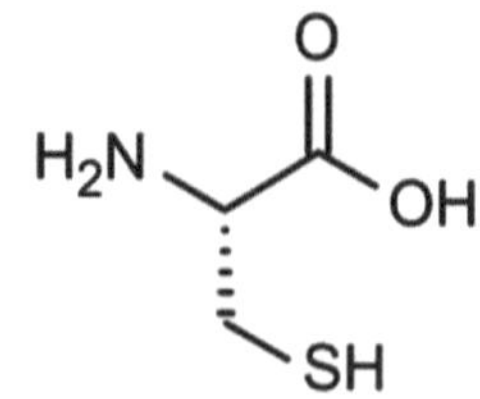

Fig. 2.3 Chemical structure of L-cysteine

$$2Ag^{+} + S^{2-} \rightarrow Ag_2S \downarrow \qquad Ksp = 8 \times 10^{-51} \qquad (2.1)$$

Consequently there is no doubt that silver would have a strong affinity with thiol groups of protein in cells, thereby inhibiting the activity of essential enzymes and finally inactivating those cells. L-cysteine is a commonly seen amino acid present in protein that contains thiol group (Fig. 2.3).

Kim and co-workers (2004) studied the synergetic effects between Ag^+ and UV-A/visible lights and they noted that this synergy was related to this silver-cysteine interaction. Briefly UV-A (or visible light) could catalyze the formation of silver-cysteine complex and this complex undergoes a photochemical reaction, producing cysteine-dimer (see Eqs. 1.1–1.3 in Chap. 1)

Second mechanism of silver ion disinfection is the interaction between ionic silver and DNA. It was reported that DNA molecules of *E. coli* became condensed after Ag^+ treatment and it was observed through TEM characterization (Feng et al. 2000). Such condensation is believed to occur naturally and it could render the incapability of DNA replication. Whether condensed DNA molecules are attributed to silver-DNA combination or it is only a defense reaction of *E. coli* still remains a question. Nonetheless, one thing is certain: silver ions have interaction with DNA molecules as reported in other literature (Rahn and Landry 1973; Rahn et al. 1973). This Ag-DNA interaction could be utilized to enhance disinfection performance: it was reported that there was a significant synergy between UV and silver ion and this synergy could be applied as a feasible disinfection strategy (Butkus et al. 2004, 2005).

Park and coworkers proposed and proved another mechanism behind aqueous silver disinfection: silver-ion-mediated reactive oxygen species (ROS) generation (Park et al. 2009). These authors used bacterial reporter strains to specifically respond to superoxide radicals (O^{2-}). Results showed that more than 50 % of log inactivation resulted from generation of O^{2-} when inactivating either *E. coli* or *Staphylococcus aureus*. This ROS mechanism well explains the fact that silver ions exhibit better disinfection performance in aerobic condition than in anaerobic condition.

2.3.2 *Antibacterial Mechanisms of Silver Nanoparticles Disinfection*

To date, antimicrobial effects of nanosilver have not yet been understood thoroughly. Many researchers proposed that releasing Ag^+ from the surface of nanoparticles may be one of mechanisms because antimicrobial effects of Ag^+ were comparable or even better than that of AgNPs (Li et al. 2008; Zodrow et al. 2009; Sintubin et al. 2011). This observation was supported by some of the cases, but there were also other studies in which nanosilver disinfection could not be simply attributed to the silver ion released from the nanoparticles (De Gusseme et al. 2009; Yin et al. 2011).

Thus far there have been limited studies that systematically discuss the intrinsic nature of nanosilver in terms of antimicrobial effects. This question remains challenging mainly due to the fact that suspension of nanosilver in water is always inevitably associated with releasing Ag^+. This point of view is in agreement with El Badawy et al. (2011). In supporting material of El Badawy et al. (2011), the authors listed 20 publications in which purification of nanosilver suspension (i.e. removing Ag^+ from nanosilver) was not explicitly mentioned. Therefore, it is difficult to investigate the roles of Ag^+ and AgNPs respectively and determine the main cause to which the observed microbial damages were responsible.

Though it was difficult to distinguish the contributions of NP from that of ions, Fabrega and co-workers managed to find some evidence for the existence of AgNPs' 'extra' effects (Fabrega et al. 2009). It was found that introducing the humic substances into the system was able to mitigate the bactericidal effect of AgNPs, yet had no impacts on toxicity of silver ions. In the presence of such humic substances, the toxicity of AgNPs could have persisted if Ag^+ had been the main mechanism. Thus comparing the distinct observations of microorganisms interacting with AgNPs and Ag^+ in the presence of humic substances could serve as a proof that Ag^+ was not the primary mechanism of AgNPs toxicity.

El Badawy and co-workers' study may provide us with some direct hints on the intrinsic antimicrobial effects of AgNPs as well (El Badawy et al. 2011). In their study, four types of silver nanoparticles, uncoated H_2-AgNPs, citrate coated AgNPs, PVP coated AgNPs and branched polyethyleneimine (BPEI) coated AgNPs, were chosen to study the relative toxicity against Bacillus species. One of the highlights of this study was using a continuous flow purification system to wash away any Ag^+ released so as to obtain simply the nanoparticles. The reduction of impurities was indicated by reduction in suspension's conductivity and further confirmed by X-ray photoelectron Spectroscopy (XPS) and proton Nuclear Magnetic Resonance (H-NMR). Reduction in toxicity was observed after the purification, which was most likely caused by the removal of Ag^+, proving Ag^+ did contribute to the bactericidal effects. This study further found that these four types of AgNPs studied were of the similar sizes (10–18 nm) but significantly different in toxicity against bacteria tested. The only possible reason for this observation was the difference in surface charge. Zeta potentials of these AgNPs were −22, −40, −12 and +39 mV, for H_2-AgNPs, citrate-AgNPs, PVP-AgNPs and BPEI-AgNPs, respectively. Considering the electrostatic interaction between negatively-charged bacteria membrane and various AgNPs, BPEI-AgNPs (+39 mV) may attract the surface of bacteria, expressing the highest toxicity. As for another three AgNPs, toxicity should be in line with electrostatic repulsive force: the most negatively-charged AgNPs has the highest repulsive electrostatic force, being the least toxic. The observed results well verified this hypothesis: citrate-AgNPs (−40 mV) showed the lowest toxicity and the positively-charged BPEI-AgNPs the highest. Moreover, this theory was in line with the study on effects of Mg^{2+} and Ca^{2+} (Jin et al. 2010) and the case of bactericidal nano-MgO (Stoimenov et al. 2002).

El Badawy's study is of great importance in understanding the intrinsic bactericidal mechanism of nanosilver. Yet this surface-related physical interaction was not fully capable of addressing some questions. For instance, if the electrostatic force was the main cause, one could expect that gold nanoparticles of positive charges would attract and inactivate any negatively-charged bacteria, such as *E. coli*. Yet majority of studies reported the opposite, that gold nanoparticles did not have any antimicrobial effects against *E. coli* (Hwang et al. 2008).

Hwang et al. (2008) applied stress-specific bioluminescent bacteria to study the toxic mode of AgNPs. These special strains were sensitive to oxidative damage, protein/membrane and DNA damages, respectively. If any above-mentioned damage occurred, corresponding gene could be induced and strongly expressed, responding bioluminescence signal. Through such experiments, the entire procedure of bactericidal effects of AgNPs was described as follows: silver ions produced from AgNPs moved inside the bacterial cell, generating superoxide radicals and other ROS via reaction with oxygen. At the same time, AgNP itself damaged the cell membrane (Morones et al. 2005) and this in turn led to the cell being unable to extrude the Ag^+, which enhanced the killing effects of Ag^+. This observation explains why AgNPs caused more damage than Ag^+ did. Hwang concluded that AgNPs (including Ag^+ released) caused toxicity via protein and membrane damage but did not induce DNA damage.

Systematic study of antibacterial activities of AgNPs was not investigated until the advent of Sintubin et al. (2011). In their study, Sintubin and co-workers investigated independently three commonly proposed mechanisms behind antibacterial activities of AgNPs, namely release of Ag^+, generation of ROS and direct contact (Sintubin et al. 2011). In order to study the contribution of Ag^+ release, minimal inhibition concentration (MIC) and minimal bactericidal concentration (MBC) and log removal of bacteria were determined and compared between AgNPs and Ag^+. Results showed that the antibacterial profiles of AgNPs and Ag^+ were the same or comparable, indicating the major contribution of bactericidal effects may be attributed to Ag^+. Impacts of ROS generation by AgNPs were investigated by comparing antibacterial activities under different O_2 percentages. This experimental design was based on one assumption that different O_2 percentage could directly affect the amount of ROS generated. If antibacterial effects mainly relied on ROS generated, different O_2 percentages would reflect on differences in antibacterial activities. Results obtained from their observation indicated that ROS generated by AgNPs most probably contributed only to a minor extent. Lastly, study on direct contact was done by measuring cell membrane permeability and by introducing dialysis membrane. Measurement of cell membrane permeability could quantify the membrane damage induced by AgNPs and introducing dialysis membrane could explicitly help us understand the difference with or without the presence of direct contact. The results obtained clearly indicated an insignificant role of direct contact in antibacterial activities of AgNPs.

In their study, Sintubin et al. (2011) also provided practical and useful experiment strategies to study the AgNP mechanism. Nevertheless, there was still room for improvement in methodology reported in their study, such as lack of study in DNA damage.

2.3.3 Antiviral Mechanisms of Silver Nanoparticles Disinfection

Compared to bacteria, viruses have huge differences that may have impacts on resistance to AgNPs and inactivation mechanisms thereof. Firstly, size of virus generally is smaller than bacterium. For instance, size of MS2 bacteriophage is approximate 25 nm while size of *E. coli* is approximate 1 micron. The size difference may lead to difference in mode of contact with AgNPs. Secondly, virus lacks cellular structure. Shell of virus only consists of protein coat and in some cases, lipid envelope surrounding the coat. In contrast, bacterial membrane has a more complicated structure, which involves lipopolysaccharides (gram-negative bacteria), peptidoglycan, protein, etc. This structure difference could also lead to difference in mode of contact with AgNPs. Thirdly, viruses do not have metabolism. Intact protein capsid and undamaged genome translates into the infectivity of virus. On the other hand, bacteria reproduction cycle requires more enzyme functions. Any intervention of these functions may lead to loss of bacterial infectivity.

Considering the huge difference between bacteria and viruses, antiviral mechanisms could be significantly different from antibacterial mechanisms. In spite of the well-documented antibacterial mechanism, systematic understanding of antiviral mechanisms of AgNPs remains a gap although some observations have been more or less reported by previous literature.

Elechiguerra et al. (2005) may be one of the earliest studies that observed interaction of AgNPs with viruses, HIV-1 specifically. Through analysis of TEM images and high angle annular dark field (HAADF) images, Elechiguerra and co-workers claimed that interaction with the gp 120 knobs of HIV was one of the major mechanisms of AgNPs to inhibit viral attachment to host cell, leading to overall inhibition of HIV infection (Elechiguerra et al. 2005). This conclusion was further confirmed by other study (Rogers et al. 2008). Similarly, TEM images of H3N2 influenza virus and adenovirus interacting with AgNPs showed that noticeable damage of the capsid was clearly proven and morphology (structure) of virus particle was breached and broken (Chen et al. 2013; Xiang et al. 2013). Furthermore, DNA copies of adenovirus decreased significantly after the AgNPs treatment in a dose-dependent manner, suggesting AgNPs are able to cause damage to gene as well. This point of view was shared by other studies (Lu et al. 2008; Lara et al. 2010). It is worth of reviewing the study of Lara et al. (2010) because the authors provided an innovative aspect of studying modes of antiviral actions by looking into the roles of AgNPs in different stages of HIV-1 life cycle. It was reported that AgNPs exerted anti-HIV effects at an early stage by inhibition of viral entry or attachment to cells. Besides, AgNPs also inhibited post-entry stages of virus life cycle, most probably by blocking HIV-1 proteins.

Compared to massive publications on antibacterial mechanisms, there have been limited studies on antiviral mechanism thus far. Furthermore, most of the studies in this part focus merely on viral damage, either capsid damage or genome damage. However, complete mechanism study must consist of studies on both AgNPs (antiviral agent) and viruses (target). Study on AgNPs is to find out what is the key property that determines the antiviral effects. Answering this could help us develop better nanosilver. On the other hand, study on viruses is to investigate what types of viral damage has been caused. Investigation of viral damage caused by AgNPs will equip us with a better understanding of virus behaviours after AgNPs treatment.

2.4 Continuous Disinfection System with Silver Nanoparticles

Many efforts have been put in application of AgNPs in continuous flow-through system. Review of previous literature shows that AgNPs has been successfully incorporated in many materials or devices to serve as disinfectants, such as polymer foam (Jain and Pradeep 2005), ceramic materials (Oyanedel-Craver and Smith 2008; Lv et al. 2009), filter paper (Dankovich and Gray 2011) and membrane (Zodrow et al. 2009; De Gusseme et al. 2011). Detailed review is presented below.

Polyurethane (PU) foams were soaked in Ag@citrate NPs suspension for coating overnight by Jain and Pradeep (2005). After saturated coating, such AgNPs-contained PU foams were washed repeatedly to remove any adsorbed ions and then air-dried. With a flow rate of 0.5 L/min, 120 L of *E. coli*-contaminated tap water (10^5 CFU/mL) were passing through the PU foams and no *E. coli* colonies were found in further enumeration. Effective as this material may be, the authors failed to answer one key question: how much AgNPs were incorporated in PU foam after soaking and washing. Lack of quantification of AgNPs in PU foam may jeopardize the reproducibility of the entire experiment. And due to the same reason, it was impossible to discuss whether or not this material was economically beneficial.

With similar 'soak-and-absorb' method, Lv and coworkers synthesized AgNPs-contained ceramic composite (Lv et al. 2009). The highlight of this study was that modification of amino groups on the ceramic surface by using 3-aminopropyltriethoxysilane was conducted to ensure a better absorbance of AgNPs. Results indicated that such ceramic composite was also able to inactivate all the initial *E. coli* (10^5 CFU/mL for 500 mL) at a flow rate of 10 mL/min. This 'soak-and-absorb' strategy was easy to control but difficult to quantify the AgNPs usage. On top of this 'soak-and-absorb' strategy, Oyanedel-Craver and

Smith (2008) provided another method to fix AgNPs onto ceramic filter: painting. Straightforward as it was, concentrate suspension of AgNPs was painted on the filter surface for economical usage of AgNPs. These two ways of fixing AgNPs on the surface did not make a significant difference in terms of disinfection performances. Over 6 log inactivation of *E. coli* were achieved for at least 90 min. Ionic silver release was monitored and the concentrations of all the samples were lower than 0.1 mg/L, which was in compliance with WHO and USEPA recommendations (WHO 2006).

Dankovich and Gray synthesized nanosilver *in situ* with the presence of filter paper in order to generate a bactericidal filter paper (Dankovich and Gray 2011). Water spiked with bacteria was vertically passing through such filter paper with simply gravity. It was noted that more than 7 log removal of *E. coli* and 3 log of *Enterococci faecalis* were achieved. The researchers also monitored the concentration of releasing Ag^+ to make sure it was not higher than 0.1 mg/L. The combination of AgNPs and a paper sheet shed some light on the development of light, cheap, portable and point-of-use water purification device.

Besides antibacterial performance, antiviral effects of AgNPs-involved continuous system were also studied. Zodrow et al. (2009) synthesized membranes with AgNPs impregnated by means of the 'phase inversion' strategy. The silver-impregnated membrane showed strong inactivation capacity against MS2 coliphage as well as *E. coli* K12, and *Pseudomonas mendocina* KR1. Another team from Belgium has also considered viral inactivation using biogenic nanosilver immobilized in polyvinylidene (PVDF) fluoride membranes (De Gusseme et al. 2011). Continuous flow system was built up with such nanosilver-contained membrane to examine the viral inactivation of UZ1 bacteriophage. With a flow rate of 0.375 L/h and a hydraulic retention time (HRT) of 1 h, the continuous flow system could achieve around 4 log removal of UZ1 after 2 h and the concentration of Ag^+ in finished water was as low as 27 ± 8 μg/L.

These studies provide invaluable experience on how to implement AgNPs in a continuous flow system and have undeniable achievements on antimicrobial effects of AgNPs. However, these studies ignore several challenges. Firstly, water matrices of real water are reported to affect the antimicrobial effects of AgNPs but not considered in those studies. Secondly, the long-term performance of such continuous flow system was not investigated. In addition, recovery methods after continuous usage need to be addressed.

References

Ahmad A, Mukherjee P, Senapati S, Mandal D, Khan MI, Kumar R, Sastry M (2003) Extracellular biosynthesis of silver nanoparticles using the fungus *Fusarium oxysporum*. Colloid Surf B 28:313–318

Baalousha M, Manciulea A, Cumberland S, Kendall K, Lead JR (2008) Aggregation and surface properties of iron oxide nanoparticles: influence of pH and natural organic matter. Environ Toxicol Chem 27:1875–1882

Basavaraja S, Balaji SD, Lagashetty A, Rajasab AH, Venkataraman A (2008) Extracellular biosynthesis of silver nanoparticles using the fungus *Fusarium semitectum*. Mater Res Bull 43:1164–1170

Bfilainsa KC, D'Souza SF (2006) Extracellular biosynthesis of silver nanoparticles using the fungus *Aspergillus fumigatus*. Colloid Surf B 47:160–164

Butkus MA, Labare MP, Starke JA, Moon K, Talbot M (2004) Use of aqueous silver to enhance inactivation of coliphage MS-2 by UV disinfection. Appl Environ Microbiol 70:2848–2853

Butkus M, Talbot M, Labare M (2005) Feasibility of the silver-UV process for drinking water disinfection. Water Res 39:4925–4932

Chen SH, Carroll DL (2002) Synthesis and characterization of truncated triangular silver nanoplates. Nano Lett 2:1003–1007

Chen N, Zheng Y, Yin J, Li X, Zheng C (2013) Inhibitory effects of silver nanoparticles against adenovirus type 3 in vitro. J Virol Methods 193:470–477

Choi O, Deng KK, Kim N-J, Ross L Jr, Surampalli RY, Hu Z (2008) The inhibitory effects of silver nanoparticles, silver ions, and silver chloride colloids on microbial growth. Water Res 42:3066–3074

Dankovich TA, Gray DG (2011) Bactericidal paper impregnated with silver nanoparticles for point-of-use water treatment. Environ Sci Technol 45:1992–1998

De Gusseme B, Sintubin L, Baert L, Thibo E, Hennebel T, Vermeulen G, Uyttendaele M, Verstraete W, Boon N (2009) Biogenic silver for disinfection of water contaminated with viruses. Appl Environ Microbiol 76:1082–1087

De Gusseme B, Hennebel T, Christiaens E, Saveyn H, Verbeken K, Fitts JP, Boon N, Verstraete W (2011) Virus disinfection in water by biogenic silver immobilized in polyvinylidene fluoride membranes. Water Res 45:1856–1864

Diegoli S, Manciulea AL, Begum S, Jones IP, Lead JR, Preece JA (2008) Interaction between manufactured gold nanoparticles and naturally occurring organic macromolecules. Sci Total Environ 402:51–61

El Badawy AM, Luxton TP, Silva RG, Scheckel KG, Suidan MT, Tolaymat TM (2010) Impact of environmental conditions (pH, ionic strength, and electrolyte type) on the surface charge and aggregation of silver nanoparticles suspensions. Environ Sci Technol 44:1260–1266

El Badawy AM, Silva RG, Morris B, Scheckel KG, Suidan MT, Tolaymat TM (2011) Surface charge-dependent toxicity of silver nanoparticles. Environ Sci Technol 45:283–287

Elechiguerra JL, Burt JL, Morones JR, Camacho-Bragado A, Gao X, Lara HH, Yacaman MJ (2005) Interaction of silver nanoparticles with HIV-1. J Nanobiotechnol 3:6

Fabrega J, Fawcett SR, Renshaw JC, Lead JR (2009) Silver nanoparticle impact on bacterial growth: effect of pH, concentration, and organic matter. Environ Sci Technol 43:7285–7290

Feng QL, Wu J, Chen GQ, Cui FZ, Kim TN, Kim JO (2000) A mechanistic study of the antibacterial effect of silver ions on *Escherichia coli* and *Staphylococcus aureus*. J Biomed Mater Res 52:662–668

Gaikwad S, Ingle A, Gade A, Rai M, Falanga A, Incoronato N, Russo L, Galdiero S, Galdiero M (2013) Antiviral activity of mycosynthesized silver nanoparticles against herpes simplex virus and human parainfluenza virus type 3. Int J Nanomed 8:4303–4314

Gong P, Li H, He X, Wang K, Hu J, Tan W, Zhang S, Yang X (2007) Preparation and antibacterial activity of Fe_3O_4@Ag nanoparticles. Nanotechnology 18:285604

Govindaraju K, Basha SK, Kumar VG, Singaravelu G (2008) Silver, gold and bimetallic nanoparticles production using single-cell protein (*Spirulina platensis*) Geitler. J Mater Sci 43:5115–5122

Gupta A, Maynes M, Silver S (1998) Effects of halides on plasmid-mediated silver resistance in *Escherichia coli*. Appl Environ Microbiol 64:5042–5045

Hwang ET, Lee JH, Chae YJ, Kim YS, Kim BC, Sang B-I, Gu MB (2008) Analysis of the toxic mode of action of silver nanoparticles using stress-specific bioluminescent bacteria. Small 4:746–750

Jain P, Pradeep T (2005) Potential of silver nanoparticle-coated polyurethane foam as an antibacterial water filter. Biotechnol Bioeng 90:59–63

Jin RC, Cao YW, Mirkin CA, Kelly KL, Schatz GC, Zheng JG (2001) Photoinduced conversion of silver nanospheres to nanoprisms. Science 294:1901–1903

Jin X, Li MH, Wang JW, Marambio-Jones C, Peng FB, Huang XF, Damoiseaux R, Hoek EMV (2010) High-throughput screening of silver nanoparticle stability and bacterial inactivation in aquatic media: influence of specific ions. Environ Sci Technol 44:7321–7328

Juibari MM, Abbasalizadeh S, Jouzani GS, Noruzi M (2011) Intensified biosynthesis of silver nanoparticles using a native extremophilic *Ureibacillus thermosphaericus* strain. Mater Lett 65:1014–1017

Kalishwaralal K, Deepak V, Ram Kumar Pandian S, Kottaisamy M, BarathmaniKanth S, Kartikeyan B, Gurunathan S (2010) Biosynthesis of silver and gold nanoparticles using *Brevibacterium casei*. Colloids Surf B Biointerfaces 77:257–262

Kim Y-J, Song JH (2014) Synthesis of multi-walled carbon nanotube-Ag-nanoparticles composite nanomaterials using proton beam irradiation. J Nanosci Nanotechnol 14:5464–5467

Kim J, Kuk E, Yu K, Park S, Lee H, Kim S, Park Y, Hwang C (2007) Antimicrobial effects of silver nanoparticles. Nanomed Nanotechnol Biol Med 3:95–101

Kowshik M, Ashtaputre S, Kharrazi S, Vogel W, Urban J, Kulkarni SK, Paknikar KM (2003) Extracellular synthesis of silver nanoparticles by a silver-tolerant yeast strain MKY3. Nanotechnology 14:95–100

Lara HH, Ayala-Nunez NV, Ixtepan-Turrent L, Rodriguez-Padilla C (2010) Mode of antiviral action of silver nanoparticles against HIV-1. J Nanobiotechnol 8(1):1

Levard C, Mitra S, Yang T, Jew AD, Badireddy AR, Lowry GV, Brown GE Jr (2013) Effect of chloride on the dissolution rate of silver nanoparticles and toxicity to *E. coli*. Environ Sci Technol 47:5738–5745

Li Q, Mahendra S, Lyon D, Brunet L, Liga M, Li D, Alvarez P (2008) Antimicrobial nanomaterials for water disinfection and microbial control: potential applications and implications. Water Res 42:4591–4602

Li S, Zhou Z, Zhang T, Jiang G, Su R (2014) Synthesis and characterization of Ag/Fe_3O_4 electromagnetic shielding particles. J Magn Magn Mater 358:27–31

Liau SY, Read DC, Pugh WJ, Furr JR, Russell AD (1997) Interaction of silver nitrate with readily identifiable groups: relationship to the antibacterial action of silver ions. Lett Appl Microbiol 25:279–283

Lu L, Sun RW-Y, Chen R, Hui C-K, Ho C-M, Luk JM, Lau GKK, Che C-M (2008) Silver nanoparticles inhibit hepatitis B virus replication. Antivir Ther 13:253–262

Lukman AI, Gong B, Marjo CE, Roessner U, Harris AT (2011) Facile synthesis, stabilization, and anti-bacterial performance of discrete Ag nanoparticles using *Medicago sativa* seed exudates. J Colloid Interface Sci 353:433–444

Lv YH, Liu H, Wang Z, Liu SJ, Hao LJ, Sang YH, Liu D, Wang JY, Boughton RI (2009) Silver nanoparticle-decorated porous ceramic composite for water treatment. J Membrane Sci 331:50–56

Majdalawieh A, Kanan MC, El-Kadri O, Kanan SM (2014) Recent advances in gold and silver nanoparticles: synthesis and applications. J Nanosci Nanotechnol 14:4757–4780

Metraux GS, Mirkin CA (2005) Rapid thermal synthesis of silver nanoprisms with chemically tailorable thickness. Adv Mater 17:412–415

Millstone JE, Hurst SJ, Métraux GS, Cutler JI, Mirkin CA (2009) Colloidal gold and silver triangular nanoprisms. Small 5:646–664

Moore MN (2006) Do nanoparticles present ecotoxicological risks for the health of the aquatic environment? Environ Int 32:967–976

Mori Y, Ono T, Miyahira Y, Vinh Quang N, Matsui T, Ishihara M (2013) Antiviral activity of silver nanoparticle/chitosan composites against H1N1 influenza A virus. Nanoscale Res Lett 8

Morones JR, Elechiguerra JL, Camacho A, Holt K, Kouri JB, Ramírez JT, Yacaman MJ (2005) The bactericidal effect of silver nanoparticles. Nanotechnology 16:2346–2353

Niemeyer CM (2001) Nanoparticles, proteins, and nucleic acids: biotechnology meets materials science. Angew Chem Int Edit 40:4128–4158

Nowack B, Krug HF, Height M (2011) 120 Years of nanosilver history: implications for policy makers. Environ Sci Technol 45:1177–1183

Oyanedel-Craver VA, Smith JA (2008) Sustainable colloidal-silver-impregnated ceramic filter for point-of-use water treatment. Environ Sci Technol 42:927–933

Pal S, Tak YK, Song JM (2007) Does the antibacterial activity of silver nanoparticles depend on the shape of the nanoparticle? A study of the gram-negative bacterium *Escherichia coli*. Appl Environ Microbiol 73:1712–1720

Panacek A, Kvitek L, Prucek R, Kolar M, Vecerova R, Pizurova N, Sharma VK, Nevecna Tj, Zboril R (2006) Silver colloid nanoparticles: synthesis, characterization, and their antibacterial activity. J Phys Chem B 110:16248–16253

Park H-J, Kim JY, Kim J, Lee J-H, Hahn J-S, Gu MB, Yoon J (2009) Silver-ion-mediated reactive oxygen species generation affecting bactericidal activity. Water Res 43:1027–1032

Park HH, Park S, Ko G, Woo K (2013) Magnetic hybrid colloids decorated with Ag nanoparticles bite away bacteria and chemisorb viruses. J Mater Chem B 1:2701–2709

Rahn RO, Landry LC (1973) Ultraviolet-irradiation of nucleic-acids complexed with heavy-atoms. 2. Phosphorescence and photodimerization of DNA complexed with Ag. Photochem Photobiol 18:29–38

Rahn RO, Setlow JK, Landry LC (1973) Ultraviolet-irradiation of nucleic-acids complexed with heavy-atoms. 3. Influence of Ag^+ and Hg^{2+} on sensitivity of phage and of transforming DNA to ultraviolet-radiation. Photochem Photobiol 18:39–41

Rajasekharreddy P, Rani PU, Sreedhar B (2010) Qualitative assessment of silver and gold nanoparticle synthesis in various plants: a photobiological approach. J Nanopart Res 12: 1711–1721

Rogers JV, Parkinson CV, Choi YW, Speshock JL, Hussain SM (2008) A preliminary assessment of silver nanoparticle inhibition of monkeypox virus plaque formation. Nanoscale Res Lett 3:129–133

Russell AD, Hugo WB (1994) Antimicrobial activity and action of silver. Prog Med Chem 31:351–370

Shrivastava S, Bera T, Roy A, Singh G, Ramachandrarao P, Dash D (2007) Characterization of enhanced antibacterial effects of novel silver nanoparticles. Nanotechnology 18:225103

Silver S, Phung LT, Silver G (2006) Silver as biocides in burn and wound dressings and bacterial resistance to silver compounds. J Ind Microbiol Biotechnol 33:627–634

Sintubin L, De Windt W, Dick J, Mast J, van der Ha D, Verstraete W, Boon N (2009) Lactic acid bacteria as reducing and capping agent for the fast and efficient production of silver nanoparticles. Appl Microbiol Biotechnol 84:741–749

Sintubin L, De Gusseme B, Van der Meeren P, Pycke BFG, Verstraete W, Boon N (2011) The antibacterial activity of biogenic silver and its mode of action. Appl Microbiol Biot 91:153–162

Sintubin L, Verstraete W, Boon N (2012) Biologically produced nanosilver: current state and future perspectives. Biotechnol Bioeng 109:2422–2436

Sondi I, Salopek-Sondi B (2004) Silver nanoparticles as antimicrobial agent: a case study on *E. coli* as a model for Gram-negative bacteria. J Colloid Interface Sci 275:177–182

Sondi I, Goia DV, Matijević E (2003) Preparation of highly concentrated stable dispersions of uniform silver nanoparticles. J Colloid Interface Sci 260:75–81

Stoimenov PK, Klinger RL, Marchin GL, Klabunde KJ (2002) Metal oxide nanoparticles as bactericidal agents. Langmuir 18:6679–6686

Sun YG, Xia YN (2002) Shape-controlled synthesis of gold and silver nanoparticles. Science 298:2176–2179

Veerasamy R, Xin TZ, Gunasagaran S, Xiang TFW, Yang EFC, Jeyakumar N, Dhanaraj SA (2011) Biosynthesis of silver nanoparticles using mangosteen leaf extract and evaluation of their antimicrobial activities. J Saudi Chem Soc 15:113–120

WHO (2006) Guidelines for drinking water quality. In: Organization WH (ed). IWA Publishing, London

Xia Y, Xiong Y, Lim B, Skrabalak SE (2008) Shape-controlled synthesis of metal nanocrystals: simple chemistry meets complex physics? Angew Chem Int Ed 48:60–103

Xiang D, Zheng Y, Duan W, Li X, Yin J, Shigdar S, O'Connor ML, Marappan M, Zhao X, Miao Y, Xiang B, Zheng C (2013) Inhibition of A/Human/Hubei/3/2005 (H3N2) influenza virus infection by silver nanoparticles in vitro and in vivo. Int J Nanomed 8:4103–4113

Yin L, Cheng Y, Espinasse B, Colman BP, Auffan M, Wiesner M, Rose J, Liu J, Bernhardt ES (2011) More than the ions: the effects of silver nanoparticles on *Lolium multiflorum*. Environ Sci Technol 45:2360–2367

Ziemkowska W, Basiak D, Kurtycz P, Jastrzebska A, Olszyna A, Kunicki A (2014) Nano-titanium oxide doped with gold, silver, and palladium—synthesis and structural characterization. Chem Pap 68:959–968

Zodrow K, Brunet L, Mahendra S, Li D, Zhang A, Li Q, Alvarez PJJ (2009) Polysulfone ultrafiltration membranes impregnated with silver nanoparticles show improved biofouling resistance and virus removal. Water Res 43:715–723

Chapter 3
Materials and Methods

3.1 Silver Nanoparticles

Chemicals used in this study were purchased from Sigma-Aldrich, US, unless otherwise specified. Deionized water (DI water) used throughout this study was 18 mΩ pure water produced by MilliQ water purification system (Milipore, US). Commercial silver nanoparticles (99.5 % pure, referred to as 'com-AgNPs') were purchased directly from Sigma-Aldrich, US and used without further purification.

3.1.1 Chemically-Synthesized Silver Nanoparticles

Chem-AgNPs were synthesized based on procedures developed by Pal et al. (2007) with minor changes. Typically, 0.5 ml of 0.01 M $NaBH_4$ was rapidly injected into solution containing 0.5 ml of 0.01 M $AgNO_3$ and 20 ml of 1 mM sodium citrate. Resulting solution was vigorously stirred for 5 min and aged for 1.5 h to serve as Ag seed solution. Thereafter, 100 ml of 1 mM $AgNO_3$ was boiled and 3 ml of Ag seed solution and 1.04 ml of 0.1 M sodium citrate were then added and kept boiled until the solution turned greenish yellow. After being cooled down to room temperature, obtained nanosilver colloids were washed with DI water three times using centrifuge (13,000×*g*, 15 min).

3.1.2 Biologically-Synthesized Silver Nanoparticles

Bio-AgNPs was synthesized according to Sintubin et al. (2009). Typically, *L. fermentum* ATCC 11976 was purchased from American Type Culture Collection (ATCC), US. It was inoculated in autoclaved Man-Rogosa-Sharpe (MRS) broth

X. Yang, *A Study on Antimicrobial Effects of Nanosilver for Drinking Water Disinfection*, Springer Theses, DOI 10.1007/978-981-10-2902-8_3

(5.21 g dissolved into 100 mL, Oxoid, UK). After 48-hour incubation at 30 °C shaker incubator (Orbital, Taiwan), biomass of *L. fermentum* was harvested and washed three times by centrifuge (5000×*g*, 10 min). After diluting to 4.6 g [as cell dry weight (CDW)] per litre, the biomass was alkalified by adding 2.56 v% of 1 M NaOH solution. 16.5 mg/ml (as Ag) of $Ag(NH_3)_2NO_3$ solution was then added into the obtained biomass to achieve a ratio of 1:4.6 for Ag:CDW. The mixture was carefully covered by aluminum foil and shaken at room temperature (28 °C) for 24 h. Bio-AgNPs was then purified using centrifuge (10,000×*g*, 15 min).

3.1.3 Characterization of Silver Nanoparticles

TEM (JEM 3010, JEOL Ltd., Japan) was used to characterize size and shape of AgNPs at 300 kV in this study. Samples were prepared by depositing a drop of freshly prepared AgNPs suspension (aqueous) on a carbon coated copper grid (formvar mesh with double carbon coating, Analytical Technologies Pte Ltd, Singapore) and air-dried at room temperature overnight. TEM test was conducted with energy-dispersive X-ray spectroscopy (EDX, INCAx-Sight, Oxford Instruments) in order to further validating the presence of AgNPs by the occurrence of Ag-specific peaks.

Inductively coupled plasma optical emission spectrometry (ICP-OES, PerkinElmer, USA) was used to quantify the concentration of AgNPs synthesized after being digested with 65 % HNO_3 and 30 % H_2O_2 in compliance with USEPA 3050B. The typical digestion process was described below: 2 ml of AgNPs suspension was added into a beaker containing 2 ml of 65 % HNO_3 and heated to boiling. Subsequently 1 ml of 30 % H_2O_2 was added. This suspension was kept boiling until it changed from dark suspension (AgNPs) into colorless and transparent solution (ionic silver). After digestion, ICP-OES test was conducted to measure the concentration of ionic silver in the solution, so that the mass concentration of AgNPs could be determined. Detection limit of ICP-OES for silver is 0.1 mg/L and silver standard solution (Sigma-Aldrich) was used as standard curve in ICP-OES.

Zeta potential (ζ, mV) and hydrodynamic diameter (HDD, nm) of nanoparticles tested were measured by Zeta Potential Analyzer (ZetaPALS, Brookhaven Instruments Corporation, US) at pH 7. The HDD was determined with a 633 nm laser source and a detection angle of 173°.

3.2 Microorganisms Cultivation and Enumeration

3.2.1 Incubation of E. Coli *and MS2 Coliphage*

Sterilization of broth and other solution involved were done by autoclave at 121 °C for 15 min throughout the entire study. *E. coli* ATCC 15597 and MS2 coliphage

ATCC 15597-B1 were chosen as models of bacteria and viruses in water and purchased directly from ATCC, US. All incubations were conducted at 37 °C and 120 rpm. A culture of *E. coli* ATCC 15597 was incubated in tryptic soy broth (TSB, 0.9 g dissolved into 30 mL, Oxoid, UK) overnight. One milliliter of such *E. coli* overnight culture was transferred and incubated in 30 ml of fresh TSB for 4 h to obtain *E. coli* at exponential phase, which was washed three times using centrifuge (2500×*g*, 10 min) before conducting inactivation experiments. Original concentration of *E. coli* was 10^6 CFU/ml.

Stock of MS2 coliphage ATCC 15597-B1 was mixed with *E. coli* of exponential phase and incubated for 20 h. As-prepared MS2 suspension was centrifuged at 3600×*g* for 10 min and the obtained supernatant was passed through a 0.22-μm sterile filter. Filtrate was then used as the MS2 stock (10^{11} PFU/ml).

3.2.2 *Enumeration of* E. Coli *and MS2 Coliphage*

Enumeration of *E. coli* relied on plate culture method in accordance to Kim et al. (2008). A series of 10-fold dilutions were performed in 0.9 % NaCl solution and 0.1 ml of at least consecutive three dilutions was spread on TSB agar plates in duplicates, incubated at 37 °C overnight (>18 h). The dilution giving the valid number of CFU ($30 < N < 300$) was averaged and used to calculate the *E. coli* log removal.

Enumeration of MS2 was done using Double Agar Layer (DAL) method in accordance to the USEPA standard method 1602 (Butkus et al. 2004). The sample was 10-fold diluted serially to the required dilution in 0.9 % NaCl solution. Zero point one millilitre of *E. coli* in log-phase and 0.5 mL of sample were transferred and mixed in 3 mL of molten top agar at 49 °C (Butkus et al. 2004). After mixing well, the molten agar was poured on to TSB plates. After agar solidification, the plates were then incubated at 37 °C overnight (>18 h). The dilution giving the valid number of PFU ($30 < N < 300$) was averaged and used to further calculate the MS2 bacteriophage log removal.

3.3 Disinfection Experiments

AgNPs suspension was mixed with washed *E. coli* (or MS2 coliphage stock) to render a 10-ml reaction system that contained 10 mg/L of AgNPs and 10^5 CFU/ml of *E. coli* (or 10^6 PFU/ml of MS2 coliphage). Such reactions were continuously stirred at room temperature and covered by aluminum foil to minimize the impact of ambient light. Samples were taken after 5-, 10-, 30- and 60-min contact time. Proper dilution and enumeration were conducted within 5 min after sampling. Inactivation was expressed as log base 10 values, which equals $\log_{10}(C_0/C_t)$. Each experiment was conducted in triplicates and error bars denoted one standard deviation.

3.4 Effects of Water Matrices

Sodium nitrate was selected to simulate ionic strength while calcium nitrate and magnesium nitrate were selected to simulate the hardness present in natural water. Sodium chloride was introduced as the source of Cl^- while humic acid as the source of TOC. These conditions were chosen because antimicrobial effects of nanosilver were reported to be affected by these conditions according to previous studies (Fabrega et al. 2009; Jin et al. 2010; El Badawy et al. 2011). Buffer solutions were used as follows: pH 5.8—phosphate (0.1 mM); pH 7.4—phosphate (0.1 mM) and pH 9.0—Borax solution (1 mM). Concentration of bio-AgNPs and contact time were fixed at 10 mg/L and 10 min, respectively, in this part of the study.

Range of water matrices to be studied is summarized in Table 3.1. This range of water matrices was carefully selected in order to cover the typical water quality of surface water (Crittenden et al. 2005; Jin et al. 2010).

As for measurement of Ca^{2+} and Mg^{2+}, ion chromatography (IC) system (LC20 Chromatography Enclosure, Dionex, Thermo Scientific, US) was used. And TOC Analyzer (TOC-L, Shimadzu, Japan) was introduced to measure the concentration of TOC. For pH value measurement, electronic pH meter (pH 700, Eutech Instruments, Singapore) was used in this study.

3.5 Mechanism Study

3.5.1 Mode of Action of Nanosilver

Typically, bio-AgNPs suspension was mixed with MS2 coliphage stock to obtain a 10-ml reaction system unless otherwise specified. Initial concentration of MS2 coliphage employed in this study was maintained at 10^6 PFU/ml. All the reactions were conducted under continuously stirred conditions at room temperature and covered by aluminum foil to minimize the impact of ambient light. Contact time was maintained at 10 min unless specified.

In order to investigate the role of released silver ions, 157.4 mg of $AgNO_3$ was dissolved into 100 ml of DI water to obtain 1000 mg Ag/L stock solution.

Table 3.1 Range of water matrices to be studied

	Range
Cl^-	1–20 ppm
Ca^{2+}/Mg^{2+}	0.94–188 ppm as $CaCO_3$
Ionic strength	0.028–5.6 mM
TOC	1–30 ppm
pH	5.8; 7.4; 9.0

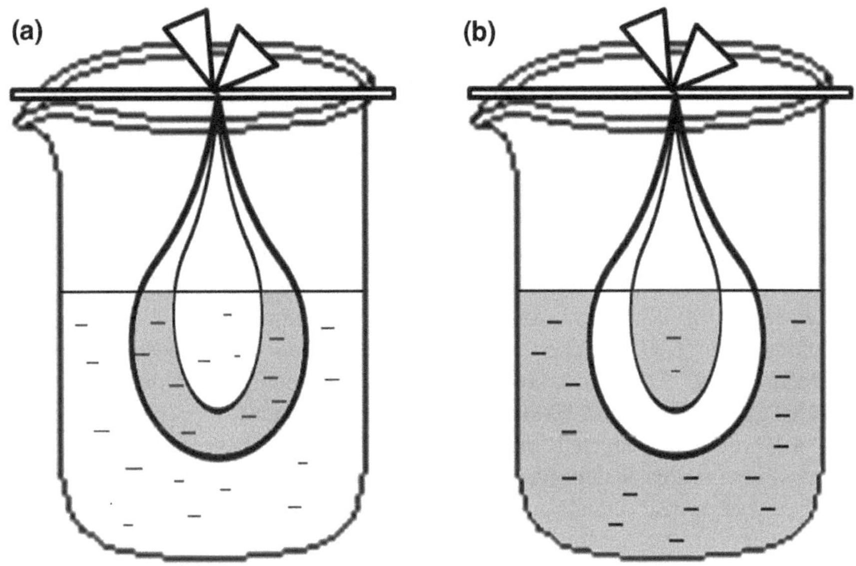

Fig. 3.1 Experiment setup to study the role of direct contact. In setup (**a**), bio-AgNPs were placed in dialysis tubing while in setup (**b**) bio-AgNPs were fully mixed with MS2 in the container. Both systems were agitated and sampled after 1-h contact time. MS2 coliphage was spiked directly in the container

Bio-AgNPs (or Ag^+) was mixed with MS2 coliphage at pre-determined concentrations (10, 25 and 50 mg/L) and contact times (5 and 10 min). Comparison of inactivation of MS2 bacteriophage was also conducted at 10 mg/L for bio-AgNPs and 0.5 mg/L for Ag^+, respectively. Samples were taken at 5, 10, 30 and 60 min.

To study the role of ROS, 15-ml penicillin bottles with 10 mg/L of bio-AgNPs and MS2 coliphage were incubated under different oxygen levels: anaerobic (0 % O_2), atmosphere (21 % O_2) and hyperoxic (60 % O_2). To establish anaerobic conditions, penicillin bottles were purged with N_2 for 5 min while 2.5 ml of pure O_2 was spiked to the headspace of the bottles to achieve hyperoxic condition. Samples under these three conditions were taken at 5-, 10-, 30- and 60-min contact time, respectively.

Role of direct contact was studied with the setups illustrated in Fig. 3.1. Dialysis membrane tubing (MWCO: 12-14,000, Spectrum Spetra/Por molecularporous membrane, Spectrum Lab, US) was used to block the direct contact between MS2 and bio-AgNPs as shown in Fig. 3.1a. Figure 3.1b was adopted as control, in which bio-AgNPs were mixed with MS2 thoroughly in the container when the same dialysis tubing was present. After 1 h agitation, samples were taken and plate cultures were conducted immediately after dilution.

3.5.2 Viral Damage

In order to study the reduction of antigenicity of MS2 coliphage, sandwich ELISA assay (MS2 virus BioThreat Alert ELISA kit pre-coated, Tetracore, US) was performed to quantify such damage according to previous study (Kim et al. 2008) as well as manufacturer's instructions. For buffer preparation, 0.8 ml of Tween-20 was added to 800 ml of PBS to obtain PBST and 5 g of dry skim milk was added into 100 ml of such PBST to gain ELISA dilution/blocking buffer. The as-received plate was blocked with 150 μl of ELISA dilution/blocking buffer at 37 °C for 1 h. Subsequently 100 μl of MS2 samples were inoculated into the well and incubated for another 1 h at 37 °C. Then, 100 μl of detector antibody (Mouse monoclonal anti-MS2 Virus, 2.5 μg/ml), 100 μl of conjugate antibody (Goat anti-mouse IgG-HRP, 0.08 μg/ml) and 100 μl of ABT 2-part peroxidase substrate were added into each well and incubated for 1 h at 37 °C, respectively, with PBST washing (4 times) between each step. Microtiter plate reader (Sunrise, Tecan, Switzerland) was then used to read the absorbance at 405 nm.

QIAamp Viral RNA Mini Kit (QIAGEN, Germany) was used to extract RNA from MS2 samples according to manufacturer's instructions. Four primer sets (Table 3.2) were selected from previous study (O'Connell et al. 2006). To perform a qRT-PCR test, iScript One-Step RT-PCR with SYBR Green kit (Bio-Rad, US) was used as received and each primer set was added to a final concentration of 400 nM in 50-μl reaction mixture. Assays were performed on iCycler iQTM real-time detection system (Bio-Rad, US) with the following protocol: 48 °C for 30 min, 95 °C for 10 min, 40 cycles of 95 °C for 15 s, 62 °C for 60 s, followed by incubations at 95 °C for 15 s, 60 °C for 15 s and 95 °C for 15 s. RNA samples were tested in duplicates and average cycle threshold (C_T) numbers were plotted against the known concentration of MS2 bacteriophage (in log value, Fig. 3.2) in

Table 3.2 Primer sets of MS2 genome used for RT-PCR test

Primer Set	Start[a]	Direction	Primer sequences (5′–3′)	Product of target gene[b]
1	1155	Forward	TGTGGAGAGACAGGGCACTG	Assembly Protein
	1231	Reverse	CAGTTGTTGGCCATACGGATT	
2	1449	Forward	CGTTCACAGGCTTACAAAGTAACCT	Coat protein
	1555	Reverse	CCAACAGTCTGGGTTGCCAC	
3	1693	Forward	CCTCAGCAATCGCAGCAAA	Lysis protein
	1807	Reverse	GGAAGATCAATACATAAAGAGTTGAACTTC	
4	2232	Forward	GCTCTGAGAGCGGCTCTATTG	RNA replicase β chain
	2301	Reverse	CGTTATAGCGGACCGCGT	

[a]Position of first base in each oligonucleotide sequence of MS2 in GenBank accession no. NC_001417

[b]Protein names as given in GenBank genome sequence accession NC_001417

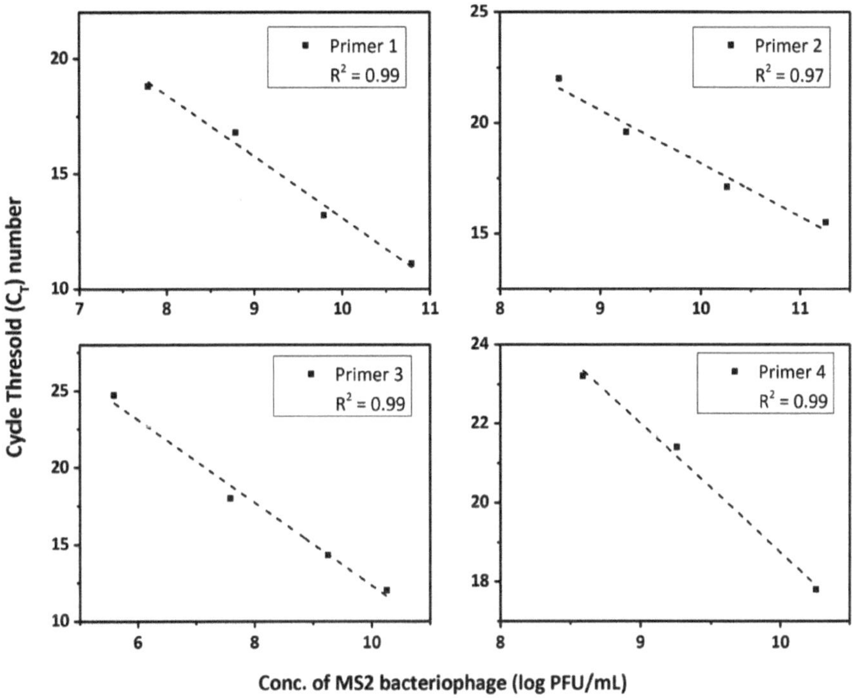

Fig. 3.2 Standard curve of 4 primer sets: the C_T number versus log value of concentration of MS2 bacteriophage

order to further calculate the log inactivation in view of genome damage (Pecson et al. 2009). Figure 3.2 clearly shows the inversely linear relationship between C_T numbers and concentrations of viable MS2 bacteriophage (in log) for all the 4 primer sets. This translated into the validity of quantifying MS2 bacteriophage using qRT-PCR.

3.6 Continuous Flow System with Nanosilver

This part of study was conducted to investigate the extent of silver nanomaterial in removing pathogens in continuous flow mode. It would be more efficient to inactivate pathogens incessantly in contaminated water once nanosilver was proved to be capable and sustainable in continuous mode. In this part, direct spray method was used to fix nanosilver in system. This method is simple and direct, focusing solely on the inherent inactivation capacity of nanosilver.

Bio-AgNPs suspension containing 5 mg of bio-AgNPs was dispersed onto 20 g of glasswool (Sigma-Aldrich, US) using spray. After drying overnight (>18 h) in oven at 50 °C, the AgNPs-containing glasswool was placed in stainless steel column (Vol = 0.22 L, Sartorius stedim, Goettingen, Germany). *E. coli* or MS2 coliphage was spiked into DI water to achieve an initial concentration of 10^7 CFU/mL or 10^6 PFU/mL, respectively, in order to mimic bacteria- or virus-contaminated water. Such synthetic water was delivered to glasswool-contained column, from bottom to top, using a peristaltic pump (6–600 rpm, Model No. 7553-75, Cole-Parmer Instrument Company) at a flow rate of 10 mL/min.

Concentrations of *E. coli* or MS2 bacteriophage in feed water (Fig. 3.3a) were deemed as initial concentration (Control). Approximate 2 ml of treated samples was collected each time directly at the outlet and were enumerated within 5 min to minimize other impacts, such as self-decay, etc. Blank experiments were also conducted with the same conditions without AgNPs coating. As for continuous flow system against *E. coli*, samples were taken at 10, 30 and 60 min. Then the whole system was stopped and backwashed at 40 mL/min of DI water for 10 min. After backwash, samples were again collected at 10, 30 and 60 min, and repeat this for the third time.

As for continuous flow system against MS2 coliphage, samples were taken at 10, 20, 30, 60, 90 and 120 min.

Backwash was implemented to investigate whether or not it could recover the disinfection performance. Glasswool column was turned upside down and DI water was delivered upwards using the same peristaltic pump at a flow rate of 40 mL/min. Duration of the backwash was 10 min each time. In the case of air-assisted backwash (air scour), air was generated by air pump (Boyu S-4000B). Subsequently the air was mixed with DI water (40 mL/min) and delivered to the column at a flow rate of 100 mL/min (Dwyer air flow meter, 100–10,000 cc/min). Photo of backwash connection was shown in Fig. 3.4.

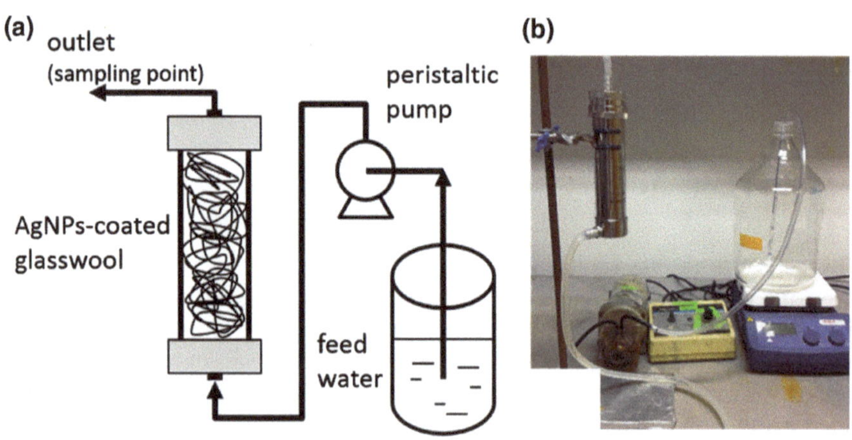

Fig. 3.3 **a** scheme of continuous flow system setup; **b** photo of continuous flow system setup

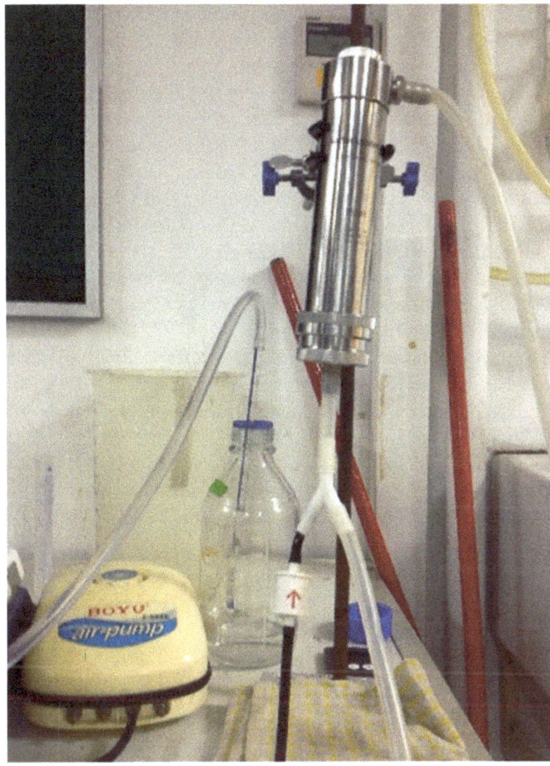

Fig. 3.4 Photo of continuous flow system during air-assisted backwash

3.7 Data Interpretation

For plate culture method, inactivation was expressed as log base 10 values, which equals $\log_{10}(C_0/C_t)$. Genome copy (presented as N_0 or N_t) was calculated using standard curve obtained by a series of concentrations of untreated MS2 as a function of CT values. Bench-scale experiments were conducted in triplicates and continuous flow experiments were repeated at least twice. Error bars denoted one standard deviation. Data were analyzed using pair-sample t-test. *P* value of 0.05 or lower was considered to be statistically significant.

References

Butkus MA, Labare MP, Starke JA, Moon K, Talbot M (2004) Use of aqueous silver to enhance inactivation of coliphage MS-2 by UV disinfection. Appl Environ Microbiol 70:2848–2853

Crittenden JC, Trussell RR, Hand DW, Howe KJ, Tchobanoglous G (2005) Water treatment: principles and design, 2nd edn. Wiley

El Badawy AM, Silva RG, Morris B, Scheckel KG, Suidan MT, Tolaymat TM (2011) Surface charge-dependent toxicity of silver nanoparticles. Environ Sci Technol 45:283–287

Fabrega J, Fawcett SR, Renshaw JC, Lead JR (2009) Silver nanoparticle impact on bacterial growth: effect of pH, concentration, and organic matter. Environ Sci Technol 43:7285–7290

Jin X, Li MH, Wang JW, Marambio-Jones C, Peng FB, Huang XF, Damoiseaux R, Hoek EMV (2010) High-throughput screening of silver nanoparticle stability and bacterial inactivation in aquatic media: influence of specific ions. Environ Sci Technol 44:7321–7328

Kim JY, Lee C, Cho M, Yoon J (2008) Enhanced inactivation of E. coli and MS-2 phage by silver ions combined with UV-A and visible light irradiation. Water Res 42:356–362

O'Connell KP, Bucher JR, Anderson PE, Cao CJ, Khan AS, Gostomski MV, Valdes JJ (2006) Real-time fluorogenic reverse transcription-PCR assays for detection of bacteriophage MS2. Appl Environ Microbiol 72:478–483

Pal S, Tak YK, Song JM (2007) Does the antibacterial activity of silver nanoparticles depend on the shape of the nanoparticle? A study of the gram-negative bacterium Escherichia coli. Appl Environ Microbiol 73:1712–1720

Pecson BM, Martin LV, Kohn T (2009) Quantitative PCR for determining the infectivity of bacteriophage MS2 upon inactivation by heat, UV-B radiation, and singlet oxygen: advantages and limitations of an enzymatic treatment to reduce false-positive results. Appl Environ Microbiol 75:5544–5554

Sintubin L, De Windt W, Dick J, Mast J, van der Ha D, Verstraete W, Boon N (2009) Lactic acid bacteria as reducing and capping agent for the fast and efficient production of silver nanoparticles. Appl Microbiol Biotechnol 84:741–749

Chapter 4
Results and Discussion

4.1 Selection of Different Nanosilver

4.1.1 Characterization of Different Nanosilver

TEM images of the nanosilver are shown in Fig. 4.1. It is noted that the com-AgNPs had severe aggregation with size ranging from 70 nm (inset of Fig. 4.1a) to several hundred nanometres (Fig. 4.1a) while chem-AgNPs had smaller size, ranging from 20 to 100 nm (Fig. 4.1b). Compared with study of Pal et al. (2007), chem-AgNPs used in this study were successfully synthesized as they had same size and shape. Bio-AgNPs (Fig. 4.1c, d) might have the best dispersion as the biomass served as scaffold, thereby avoiding aggregation from happening (Sintubin et al. 2009). Black dots seen in Fig. 4.1c, d were silver nanoparticles verified by EDX. Dark area shown in Fig. 4.1c was simply TEM copper grid determined by EDX. From Fig. 4.1c, it was clearly seen that most of silver nanoparticles were located in the cytoplasm and some of them on the cell wall (Fig. 4.1d). Apart from better dispersion, size of the bio-AgNPs was smaller (around 10 nm as shown Fig. 4.1d) than that of either com-AgNPs or chem-AgNPs. These TEM images were consistent with those presented in the study of Sintubin et al. (2009).

HDD and zeta potential of three AgNPs are shown in Table 4.1. HDD of com-AgNPs (322.7 nm) and chem-AgNPs (122.2 nm) were in good agreement with what were observed in TEM images (Fig. 4.1): com-AgNPs had the most severe aggregation while chem-AgNPs had also aggregation to some extent. It should be noted that HDD of bio-AgNPs (666.7 nm) may not be an accurate estimation of the size of bio-AgNPs. As for bio-AgNPs, silver nanoparticles were attached to the biomass of dead bacteria; however, HDD relied on laser reflection, which would treat one dead bacterium as one single particle. In fact, this explanation was supported by the fact that size of bacteria is indeed 0.5–5.0 microns, which was closed to the observed HDD (666.7 nm).

X. Yang, *A Study on Antimicrobial Effects of Nanosilver for Drinking Water Disinfection*, Springer Theses, DOI 10.1007/978-981-10-2902-8_4

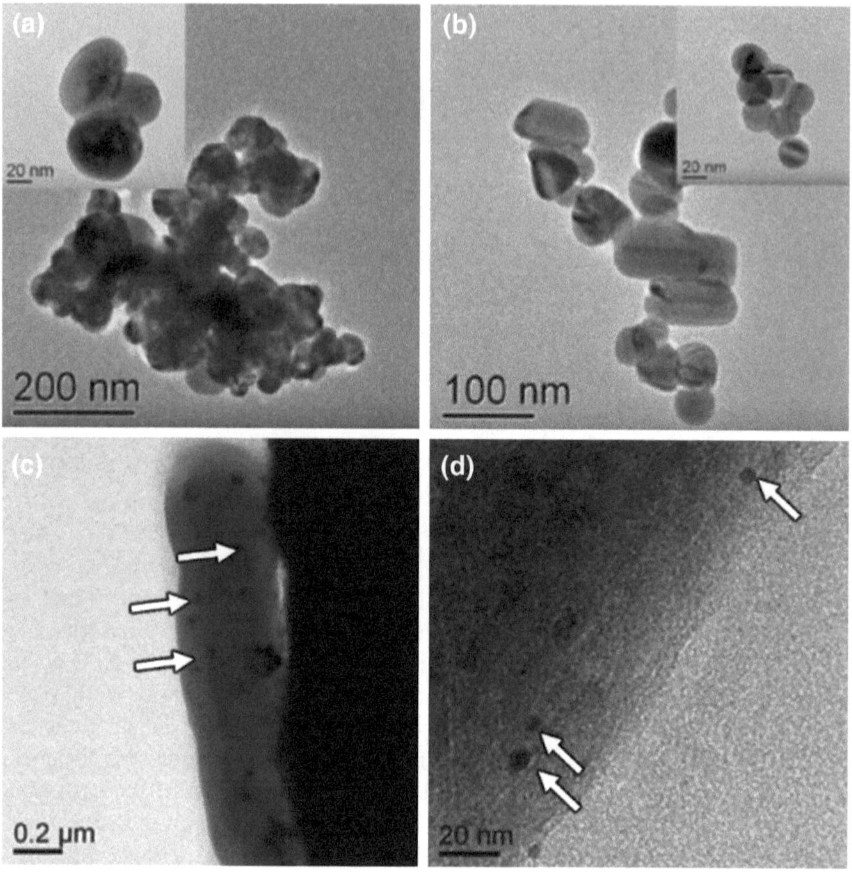

Fig. 4.1 TEM images of **a** com-AgNPs, **b** chem-AgNPs and **c**, **d** bio-AgNPs. Insets are respective single particle with the same scale bar of 20 nm

Table 4.1 HDD and zeta potential of com-AgNPs, chem-AgNPs and bio-AgNPs

	com-AgNPs	chem-AgNPs	bio-AgNPs
HDD/nm	322.7 ± 12.2	122.2 ± 1.1	666.7 ± 56.2
Zeta potential/mV	−44.32 ± 0.74	−9.06 ± 2.04	−31.96 ± 2.18

The exact size of bio-AgNPs could only be measured by TEM images. Figure 4.2 is the histogram of size distribution of bio-AgNPs determined by multiple TEM images (Figure A.1 in Appendix). The effective size of bio-AgNPs shown here was approximately 30 nm, the smallest material among these three.

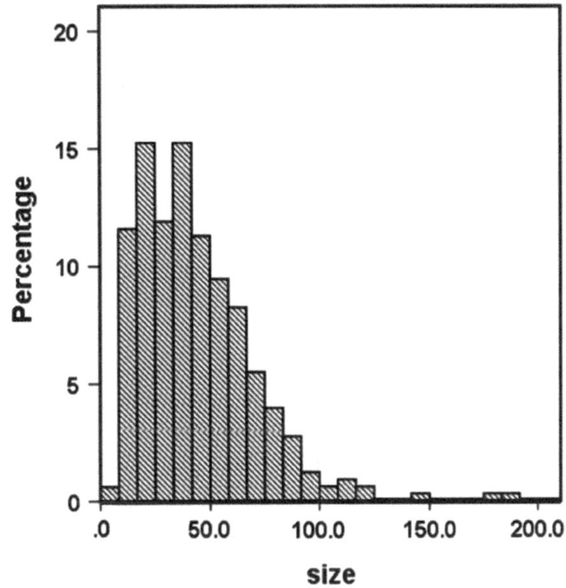

Fig. 4.2 Histogram of size distribution of bio-AgNPs: percentage versus size (nm)

4.1.2 *Comparison of Effects of Different Nanosilver Against* E. Coli

The profiles of log removal of *E. coli* versus contact time associated with com-AgNPs, chem-AgNPs and bio-AgNPs against *E. coli* are presented in Fig. 4.3. It is noted that performances of com-AgNPs and chem-AgNPs were gradually enhanced by the increase of contact time from 5 min to 60 min; however, no significant difference between these performances were detected ($P > 0.05$). One

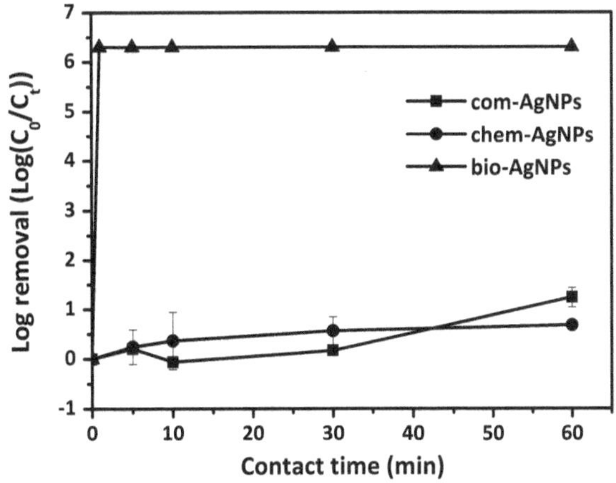

Fig. 4.3 Comparison of log removals of *E. coli* by three types of nanosilver. Concentration of AgNPs was 10 mg/L. Some error bars may be smaller than the depicted symbols

possible reason why chem-AgNPs was no better than com-AgNPs could be attributed to similar severe aggregation of nanoparticles as characterized previously. This explanation was in good agreement with previous study regarding the size effect of nanosilver (Morones et al. 2005): the nanoparticles that contributed the most were those smaller ones. On the other hand, excellent disinfection performances were achieved by bio-AgNPs. There were no colonies seen on the plates after overnight incubation of bio-AgNPs treated samples, even for those samples with 5-min contact time, which indicated a greater than 6.3 log removal. In fact, same observations were derived from experiments with shorter (1-min) and longer (10-, 30- and 60-min) contact time. Considering this 6.3-log removal is the detection limit of this method, no error bar was given.

The conclusion that bio-AgNPs had much better performance against *E. coli* was in good agreement with Sintubin et al. (2011). It was reported that bio-AgNPs have much lower values of MIC and MBC compared with the corresponding values associated with chem-AgNPs (Sintubin et al. 2011): MIC and MBC for bio-AgNPs were 12.5 mg/L and 25 mg/L, respectively while MIC and MBC for chem-AgNPs were both 500 mg/L. The authors also attributed such difference to the size effect (Morones et al. 2005; Lok et al. 2007; Pal et al. 2007). In spite of the different quantification methods (MIC/MBC versus plate culture method), their findings agree with the observation obtained in this study that bio-AgNPs were more likely to be the better disinfectant compared with com-AgNPs and chem-AgNPs.

4.1.3 Comparison of Effects of Different Nanosilver Against MS2 Coliphage

The performances of three types of nanosilver in terms of MS2 coliphage inactivation are shown in Fig. 4.4. Being similar to the results of *E. coli*, bio-AgNPs achieved the best performance among the three, inactivating 3.0 and 4.3 log MS2 with contact time of 5 and 10 min, respectively. Extending contact time to 30 min or even 60 min did not increase the log removal much. On the other hand, com-AgNPs and chem-AgNPs were not effective in inactivating MS2 bacteriophage. Comparing the performances of bio-AgNPs in Figs. 4.3 and 4.4, more than 6 log removal of *E. coli* and more than 4 log removal of MS2 bacteriophage were achieved under the same conditions (10 mg/L of bio-AgNPs for 10 min). Possible reason for the difference in performance might be that MS2 coliphage was more resistant to bio-AgNPs than *E. coli*.

The difference among these AgNPs could be well explained by the size difference. As characterized previously, bio-AgNPs had the smallest size (20–40 nm), thereby the largest surface area among these three AgNPs (Fig. 4.1). Generally larger surface area provides more contacts between viruses and AgNPs, leading to greater viral inactivation. Section 4.3.1 will demonstrate the importance of direct contact between virus and AgNPs in terms of antiviral activities.

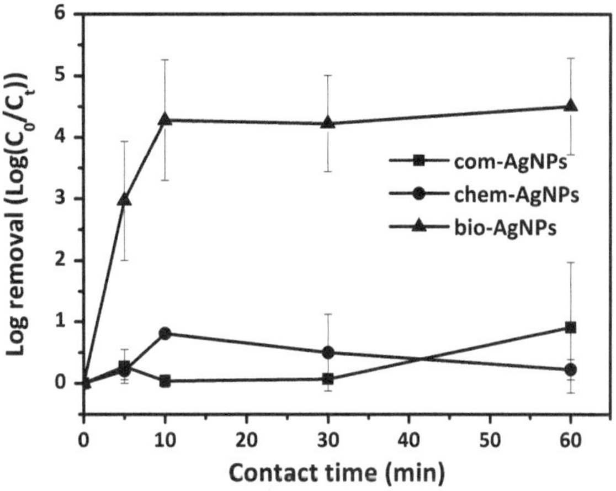

Fig. 4.4 Comparison of log removal of MS2 coliphage by three types of nanosilver. Concentration of AgNPs was 10 mg/L. Some error bars may be smaller than the depicted symbols

By comparing the antibacterial and antiviral effects of three AgNPs above, it is clearly shown that bio-AgNPs achieved the best performance. Their difference may be mainly due to the fact that bio-AgNPs had the smaller particle size than the other two. This finding was in good agreement with Morones et al. (2005) that antimicrobial effects of nanosilver were mainly contributed by small AgNPs. Therefore, bio-AgNPs were chosen to be studied in the following sections.

4.1.4 Viral Inactivation Profile of Biologically-Synthesized Silver Nanoparticles

Inactivation of MS2 coliphage by bio-AgNPs with varying contact time and concentration of bio-AgNPs was profiled in order to optimize the conditions so as to fulfil the 4-log viral removal recommended in drinking water guidelines by USEPA (Fig. 4.5). Five mg/L of bio-AgNPs was not sufficient to achieve 4 log removals even if the contact time was increased up to 120 min, while 10 mg/L of bio-AgNPs was capable of fulfilling 4-log removal with only 10 min of contact time. On the other hand, increasing dosage from 10 to 25 mg/L had no significant benefits ($P = 0.84$); and increasing the contact time did not result in an increase in log inactivation of MS2 after 10 min for 10 mg/L dosage. Such phenomenon was also observed by De Gusseme et al. (2009), in which no increase of log inactivation gained after 3 h. These two findings suggested that 10 mg/L of bio-AgNPs with 10 min of contact time represent an effective operating condition.

De Gusseme et al. (2009) reported viral removal of bio-AgNPs on UZ1 bacteriophage. They reported a slightly better disinfection performance than what was observed in this study and the main reason might be the resistance difference between UZ1 and MS2. In our understanding, UZ1, a DNA phage, was barely used

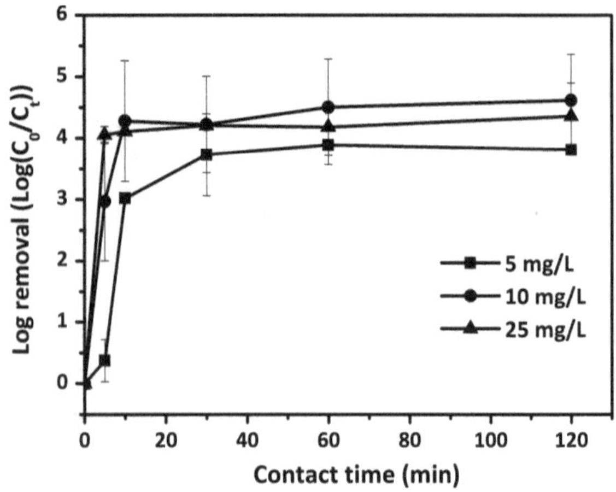

Fig. 4.5 Log removal of MS2 coliphage by bio-AgNPs

as indicator in drinking water quality control whilst previous study has already suggested that only F-specific RNA phages (e.g. MS2 coliphage) would have a resistance high enough to be considered as an acceptable indicator (Havelaar et al. 1991). Therefore study on removal of MS2 would be of greater significance in drinking water quality control.

4.2 Effects of Water Matrices

Effects of water matrices on antiviral effects of bio-AgNPs were shown and discussed in this section. The purpose of this part is to evaluate the feasibility of application of AgNPs in real water matrices. Real water matrices can be complex and unstable from time to time. Therefore in this section, several crucial parameters, namely chloride, hardness, ionic strength, TOC and pH, were introduced based on literature review. These parameters were evaluated separately in order to obtain a clearer picture.

4.2.1 Effects of Chloride

Effects of chloride on antiviral effects of bio-AgNPs were shown in Fig. 4.6. Dramatic decrease of log removal occurred with the increase of concentration of Cl^-, in which even 1 mg/L of Cl^- could decrease log removal from 4.3 to 2.5 (compare Fig. 4.6 to Fig. 4.5) and 5 mg/L or more Cl^- would totally inhibit the bio-AgNPs' performances.

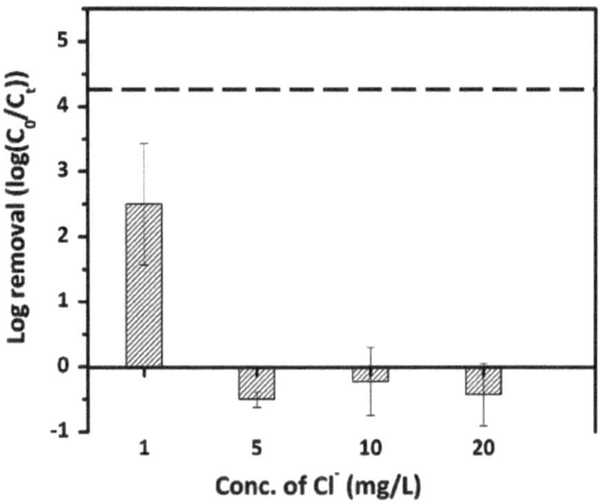

Fig. 4.6 Effect of chloride ion on log removal of MS2 coliphage by bio-AgNPs. Dashed line denotes log removal by bio-AgNPs at the same conditions in DI water

The observation regarding effects of Cl^- could be well explained by the fact that silver's toxicity largely depends on the presence of ligands (Ratte 1999). It is proposed that when Ag^+ is released from nanoparticles, Cl^- may precipitate Ag^+ in situ on the surface of nanoparticles. Such precipitate may likely form an inert layer on the surface of nanosilver that in turn 'quenched' the disinfection performance of nanosilver. This proposed explanation was in good agreement with the findings reported in previous studies (Ho et al. 2010; Levard et al. 2011).

4.2.2 *Effects of Ca/Mg/Ionic Strength*

Ca^{2+} and Mg^{2+} were reported to be able to affect silver nanoparticle stability that in turn will affect the bacterial inactivation as well (Jin et al. 2010). It was proposed by Jin et al. (2010) that Ca^{2+} and Mg^{2+} ions may serve as ion-bridges and enhance the contact between cell surface and silver nanoparticles, which were both negatively charged in their cases. Such enhanced contact would apparently enhance antibacterial effect of silver nanoparticles as observed by Jin and her co-workers. This reported enhancement would be verified in this section.

Effects of Ca^{2+} and Mg^{2+} concentrations on viral inactivation are illustrated in Fig. 4.7 (upper X-axis), which showed that log removal of MS2 coliphage decreased as the concentration of hardness increased. Log inactivation by bio-AgNPs decreased incrementally from 4.0 and 4.5 log to 1.5 and 1.6 log for Ca^{2+} and Mg^{2+}, respectively when the hardness (either Ca^{2+} or Mg^{2+}) concentration increased from 0.94 to 188 mg/L as $CaCO_3$. This observation was contradictory to Jin's observation (Jin et al. 2010). Therefore it is reasonable to believe that 'ion-bridge theory' proposed by Jin's team may not be applicable to the case of MS2, even if MS2 coliphage and bio-AgNPs were both negatively charged. One

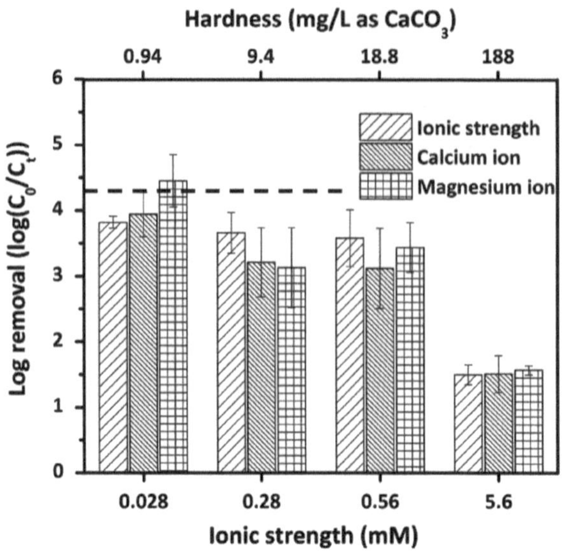

Fig. 4.7 Effects of ionic strength and hardness on log removal of MS2 coliphage by bio-AgNPs. Dashed line denotes log removal by bio-AgNPs in DI water

possible reason could be attributed to the difference between bacteria and viruses. Floyd and Sharp found that viruses tended to form aggregation regardless of the overall charge of virus particles (Floyd and Sharp 1978). Thus MS2 bacteriophage aggregated when the concentration of hardness increased. And such virus aggregate is able to protect the inner viruses and establish greater resistance to disinfectant (Mattle et al. 2011).

To further examine this result, sodium nitrate was introduced to simulate the ionic strength in previous systems, with attempt to understand whether the decreased log removals were attributed to increased ionic strength. Concentrations of $NaNO_3$ were delicately tuned so as to be equivalent to corresponding ionic strength of each hardness sample (either Ca^{2+} or Mg^{2+}). Results of $NaNO_3$ (Fig. 4.7, lower X-axis) showed that similar trend was observed and these three arrays of results, furthermore, were not significantly different ($P > 0.5$). Therefore, the adverse impacts of Ca^{2+} and Mg^{2+} might be attributed to the impacts of ionic strength and it further suggested that divalent cations might not serve as ion-bridge in the case of viral inactivation by bio-AgNPs as suggested by Jin et al. (2010).

'Ion-bridge' effects reported in the study of Jin et al. (2010) were based on IC_{50} measurements, for which 24-hr incubation with AgNPs was needed. However, the experiments conducted in this study only allowed 10-min contact time between AgNPs and MS2. 'Ion-bridge' effects, therefore, may be overshadowed by effects of ionic strength under this short contact period. As discussed previously in this section, this decreased log removal could be due to the aggregation of MS2 coliphage particles induced by high ionic strength and subsequently such aggregation would decrease the effective contact between MS2 and AgNPs (Mattle et al. 2011).

To sum up, Ca^{2+}/Mg^{2+} and ionic strength indeed had adverse impacts on antiviral effects of AgNPs. However, the impacts were less significant than chloride. It was reported that there was 5.6 mM of ionic strength in fresh river water (Jin et al. 2010). If this water was treated by a series of coagulation/flocculation and sand filtration, hardness or ionic strength would be much less than 5.6 mM, and disinfection by bio-AgNPs may become acceptable.

4.2.3 Effects of TOC

TOC may be the most important factor in drinking water treatment process and it would fluctuate on seasonal basis. The following study was to investigate whether TOC would enhance or inhibit the antiviral effects of bio-AgNPs.

Results obtained indicated that TOC had detrimental impacts on bio-AgNPs in terms of viral inactivation (Fig. 4.8). Ten milligrams per liter of TOC rendered a decrease of log removal from 3.3- to 0.4-log while further increasing concentration of TOC to 30 mg/L would totally inhibit the viral inactivation of bio-AgNPs (Fig. 4.8).

Although viral and bacterial inactivation are two different scenarios, hints might be found in the study by Fabrega et al. (2009), in which effects of organic matter was studied in the case of silver nanoparticles' impact on bacterial growth. Fabrega and her co-workers proposed that humic acid may present a physical battier between AgNPs and bacteria, thereby impeding interaction between them. Thus the results obtained in this study were in good agreement with the proposed mechanism (Fig. 4.8) and may suggest viral inactivation of bio-AgNPs might be attributed to physical contact between AgNPs and viruses.

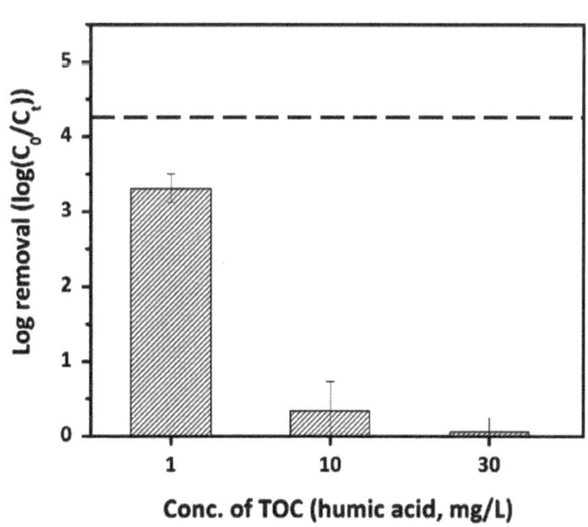

Fig. 4.8 Effects of TOC on log removal of MS2 coliphage by bio-AgNPs. Dashed line denotes log removal by bio-AgNPs at the same conditions in DI water

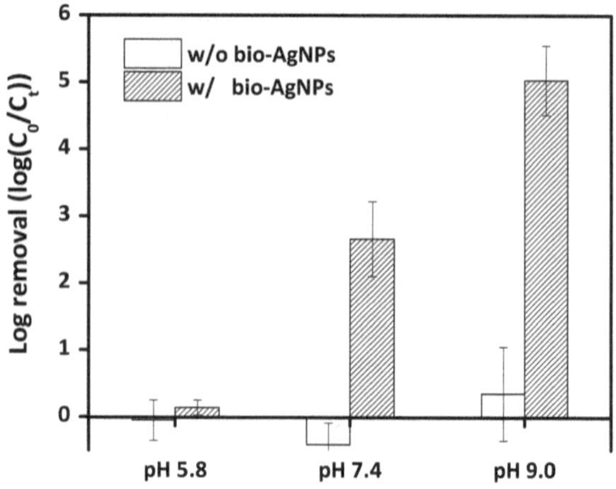

Fig. 4.9 Effects of pH on log removal of MS2 coliphage by bio-AgNPs. Dashed line denotes log removal by bio-AgNPs at the same conditions in DI water

4.2.4 *Effects of pH*

Effects of pH on removal of MS2 coliphage is shown in Fig. 4.9. It is noted that all three buffer solutions selected did not affect the MS2 infection. All the log removals with absence of bio-AgNPs (white bars) were less than 0.5-log. However, antiviral activities of bio-AgNPs were enhanced with increase of pH, from nearly no removal at pH 5.8 to greater than 4-log removal at pH 9. This pH-dependent inactivation might be due to improved release of Ag^+. Similar observation was reported in the study by Fabrega et al. (2009), in which higher pH (9.0) rendered more silver ion release, leading to greater inhibition of planktonic *P. fluorescens* cultures.

4.3 Mechanism Study of Viral Inactivation

4.3.1 *Mode of Action of Nanosilver*

To study the role of released silver ions, the most direct way is to compare the disinfection performances of Ag^+ and nanosilver under the same conditions. Figure 4.10a shows significant difference between bio-AgNPs and equivalent Ag^+ in terms of MS2 inactivation. Under all circumstances, bio-AgNPs achieved better performance than Ag^+, regardless of contact time (5 and 10 min) or mass concentration (10, 25 and 50 mg/L). This result directly showed that if all the bio-AgNPs somehow 'transformed' into Ag^+, the disinfection performance would be inhibited, suggesting that Ag^+ was not the main cause of antiviral effects of bio-AgNPs.

In fact, 10 mg/L of bio-AgNPs only released 0.5 mg/L of Ag^+ (Sintubin et al. 2011). Therefore, Fig. 4.10b presents the direct comparison of bio-AgNPs (10 mg/L) and Ag^+ (0.5 mg/L). Unsurprisingly, bio-AgNPs was able to achieve

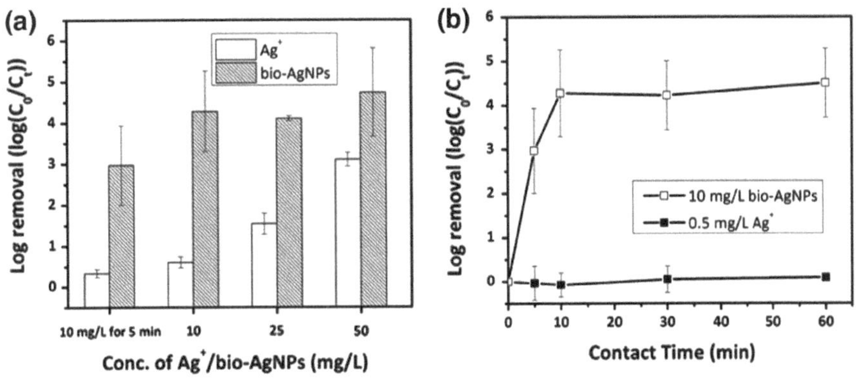

Fig. 4.10 **a** Log removal of MS2 coliphage by bio-AgNPs and equivalent Ag^+ under same conditions. **b** Comparison of log removals by 10 mg/L of bio-AgNPs and 0.5 mg/L of Ag^+

steady and substantial disinfection performance (>4-log removal) while 0.5 mg/L of Ag^+ could hardly have any antiviral effects (<0.1-log removal). This direct comparison explicitly showed that Ag^+ released from bio-AgNPs was not the main cause of antiviral effects.

Release of Ag^+ was believed to be one of the mechanisms of disinfection of microorganisms by nanosilver reported in previous studies (Matsumura et al. 2003; Sondi and Salopek-Sondi 2004; Sintubin et al. 2011). According to those existing studies, Ag^+ plays an important role in three ways: (1) combines with thiol group of several proteins, thereby resulting in malfunction of enzymes; (2) produces ROS, leading to irreversible oxidative damage to cell; and (3) possibly interacts with nucleic acid. Such conclusions were made mainly because performances of Ag^+ and nanosilver were found comparable in terms of bactericidal effects (Sondi and Salopek-Sondi 2004). However the results obtained in this study did not agree with such conclusion. There was a huge discrepancy between bio-AgNPs and equivalent Ag^+ (Fig. 4.10a), suggesting Ag^+ did not contribute much in viral inactivation. In fact this phenomenon could be easily understood considering the structural difference between bacteria and viruses. In the case of bacteria, the Ag^+ mechanism more or less depended on the fact that Ag^+ was able to enter bacterial cell, in which ion channels in cell membrane may facilitate such penetration. But none of the studies reported 'ion channel' in virion nor the penetration of any ion into a virion particle. Thus it was more likely that Ag^+ might be unable to enter coliphage and cause damage from within. On top of this, the intrinsic nature of virus does not require as many proteins as bacteria to support its functionality, rendering it less dependent on protein function and generally more resistant to Ag^+ than bacteria. Either way, Ag^+ released from nanosilver was less important in viral inactivation as observed in this study. Similar conclusion could also be found elsewhere (De Gusseme et al. 2009; Yin et al. 2011).

One assumption regarding ROS is that the greater percentage of oxygen available would induce more ROS, given that all the other conditions are the same

(Sondi and Salopek-Sondi 2004). Experiments based on this were designed to evaluate the disinfection performance under different oxygen percentages, or in other words, different concentrations of ROS. Results shown in Fig. 4.11 clearly suggested that there was no significant difference ($P > 0.005$) among anaerobic (0 % O_2), aerobic (21 % O_2) and hyperoxic (60 % O_2) conditions. Higher percentage of oxygen (60 %) was not able to enhance the antiviral effects to a large extent; neither would anaerobic condition inhibit such activities.

ROS generation was always associated with Ag^+ according to the implication of previous studies (Hwang et al. 2008; Park et al. 2009). Since contribution of Ag^+ was only limited as mentioned above, role of ROS was not predicted to be the main reason of antiviral action of nanosilver. The results obtained in this study were also in agreement with such deduction. Results obtained in this study clearly demonstrated that there was no significant difference ($P > 0.05$) when treated with bio-AgNPs under anaerobic (0 % O_2), aerobic (21 % O_2) and hyperoxic (60 % O_2) conditions. Log removals were not positively correlated to ROS generation, which suggested ROS was not the primary factor in viral inactivation by bio-AgNPs. In other words, results obtained in this study did not conflict with the fact that ROS was capable of inactivating MS2 coliphage (Kohn and Nelson 2007; Romero et al. 2011). The reason of different conclusions may be attributed to the amount of ROS produced in the system. For instance, Kohn and Nelson (2007) reported MS2 inactivation required 1 h to achieve over 4 log removal even with direct dosage of H_2O_2 (200 μM) to facilitate the ROS production. However, reaction time adopted in this study (10 min) may not be long enough to generate sufficient ROS nor to observe its contribution either. Agreement among both previous and current study may further imply that mode of actions of AgNPs in antiviral activities may depend on the way the experiments were designed and the scale of interests. For instance, high concentration of AgNPs had high probabilities to contact with virus; contribution of other mechanism (slow release of Ag^+ and generation of ROS) may

Fig. 4.11 Log removals of MS2 coliphage by 10 mg/L of bio-AgNPs under different oxygen percentage

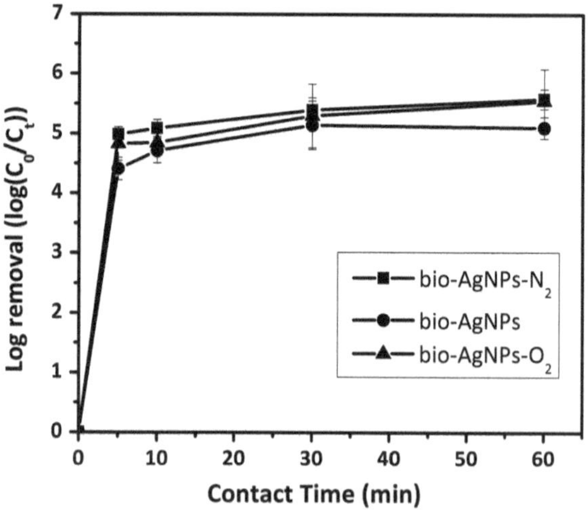

therefore be overshadowed. But when low concentration of AgNPs was used, the antiviral effects may rely on the relatively slow mechanism (slow release of Ag^+ and generation of ROS). Future work needs to done to elucidate this scale-dependent mechanism.

Study of role of direct contact was conducted using the setups shown in Fig. 3.1. Contact between bio-AgNPs and MS2 coliphage was impaired by dialysis membrane shown in Fig. 3.1a while in the other setup, MS2 coliphage was able to directly interact with bio-AgNPs in the presence of dialysis membrane as a blank control shown in Fig. 3.1b. Contact times adopted in these experiments were prolonged up to 60 min to allow sufficient time for any potential diffusion of Ag^+ or ROS between the dialysis membranes to take place. Even as the contact time increased, antiviral effects without direct contact could only show 1 log of MS2 coliphage inactivation as shown in Fig. 4.12, whereas direct contact could achieve over 4.5-log removal.

Direct physical contact of viruses with nanosilver was the last proposed mechanism to be evaluated by isolating bio-AgNPs behind dialysis membrane. Results obtained showed that with the inhibition of direct contact, viral inactivation of bio-AgNPs would decrease dramatically from 4.5- to only 1-log removal. It is still of interest to question what other factors could account for the remaining 1-log removal, other than the effects of Ag^+ and ROS. In order to answer this question, we tested and found that infectious MS2 coliphage present inside the dialysis membrane ($\sim 10^2$ PFU/mL), which suggested that MS2 coliphage was able to penetrate through the 14,000 MWCO membrane via diffusion. This finding might explain the 1-log removal: MS2 may diffuse into the dialysis tubing and then be inactivated by nanosilver. Sintubin et al. (2011) studied the mode of action of biogenic silver against bacteria. However, their study revealed a different pattern, in which the Ag^+ release was the main contributor while the effect of direct contact was only negligible in bactericidal actions. To the best of our knowledge, such difference in conclusions may be attributed to the resistance difference of bacteria

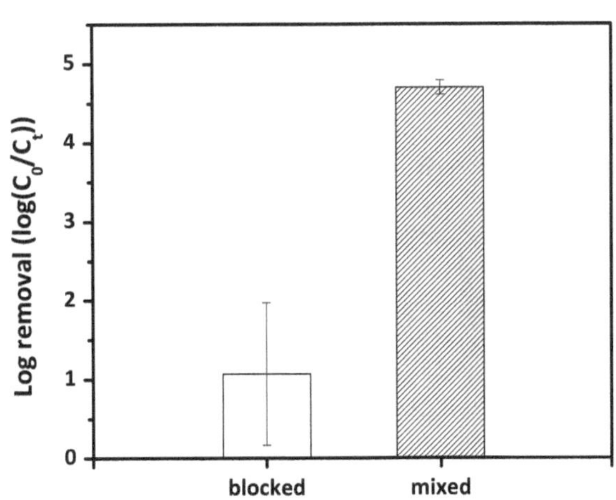

Fig. 4.12 Log removals of MS2 coliphage by bio-AgNPs when direct contact between these two was impaired ('blocked') and when these two were mixed ('mixed')

and viruses to bio-AgNPs. Our previous study experimentally confirmed such difference of resistance: *E. coli* could be fully inactivated (>6-log removal) within 1 min while only 4 log MS2 could be inactivated after 10-min treatment with same dosage of bio-AgNPs (Figs. 4.3 and 4.4).

Elechiguerra et al. (2005) visually observed silver nanoparticle physically attaching to HIV-1 using HAADF. The authors proposed that the silver nanoparticles selectively bound with gp 120 subunit of the viral envelop glycoprotein, rendering the inhibition of virus infectivity as observed. Such findings again confirmed the crucial role of direct contact in antiviral effects and more importantly, were in good agreement with our ELISA assays on antigenicity of MS2 coliphage (data will be presented in next section). ELISA assay on antigenicity has been used to examine how an intact capsid of MS2 was damaged (Kim et al. 2011). Therefore loss of antigenicity observed in this study supported one hypothesis that bio-AgNPs inactivated MS2 mainly by contacting the protein coat of viruses and rendering capsid damage, for instance antigenicity in this case, on MS2. One may argue that such reduction of absorbance may not result from impaired antigenicity of MS2; it could be that bio-AgNPs ubiquitously blocks the capture antibody considering that bio-AgNPs were a mixture of silver nanoparticles and biomass (Sintubin et al. 2009). To eliminate such possibility, negative control with chem-AgNPs and biomass, respectively, were also conducted. Compared to bio-AgNPs, chem-AgNPs and biomass could only reduce the antigenicity by 33.6 %. Our previous study affirmed that chem-Ag and biomass had little effects on MS2. Therefore, such reduction of antigenicity indicated that silver nanoparticles and biomass did more or less block capture antibodies, resulting in the observed false positive. At the same time the results of negative control demonstrated that bio-AgNPs, in spite of the blocking effects, did induce significant damage on the antigenicity.

4.3.2 Viral Damage Caused by Nanosilver

Presented in Fig. 4.13 is the reduction of MS2 antigenicity after treated with bio-AgNPs, chem-AgNPs and biomass. After treated with 10 mg/L of bio-AgNPs for 10 min, ELISA absorbance at 405 nm dropped to zero (slightly negative), indicating antigenicity of MS2 was impaired to a significantly large extent. To eliminate the possibility that the biomass in bio-AgNPs or nanoparticles themselves may induce false positive results, biomass and chem-AgNPs were also assessed in ELISA under the same conditions. These two were unable to inactivate MS2 coliphage under the conditions tested according to our previous result (Fig. 4.4). Assays with these two showed that absorbance (0.77 for chem-AgNPs and 0.80 for biomass) did drop to some extent but not as significantly as that of bio-AgNPs.

The qRT-PCR assays of MS2 samples after nanosilver treatment are summarized in Fig. 4.14 for all the four primer sets chosen. Firstly, results comparing among the 4 primer segments were insignificantly different ($P > 0.05$), illustrating that damage

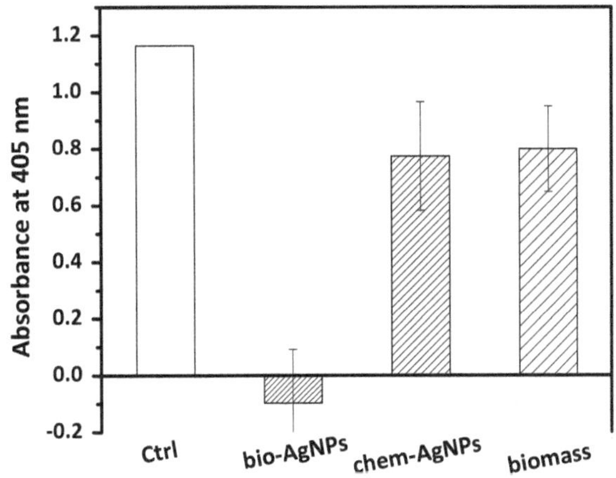

Fig. 4.13 ELISA analysis of bio-AgNPs, chem-AgNPs and biomass

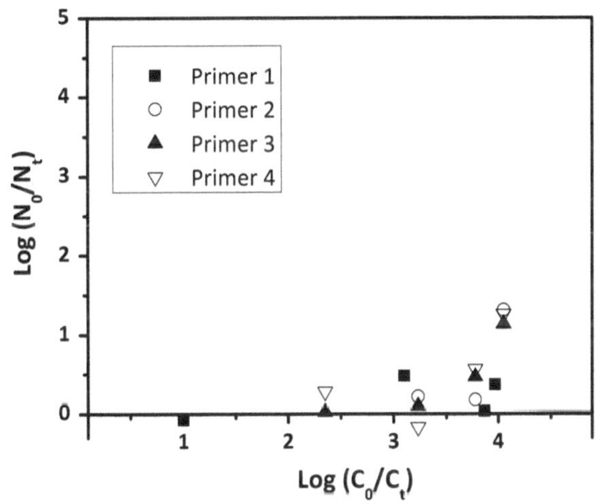

Fig. 4.14 Decay of genome (log (N_0/N_t)) as a function of decay of viral infectivity (log (C_0/C_t)) for all the 4 primer sets

caused by nanosilver was not selective with regard to genome positions. Secondly, modest genome damage of MS2 occurred following AgNPs treatment for all the 4 primer segments. When decay of infectivity (N_0/N_t) of MS2 was lower than 3 log, the genome damage (C_0/C_t) was negligible (<0.5 log). When decay of infectivity of MS2 achieved approximate 4-log removal, there was a slight increase of genome damage ($\sim$ 1.2 log).

PCR tests showed that genome decay was not relevant to infectivity decay, indicating bio-AgNPs was unlikely to cause damage on MS2 RNA. Comparison using the 4 primer sets showed that none of the genomic regions were particularly susceptible to nanosilver treatment. Such results further indicated that damage to RNA by nanosilver was not significant under the conditions investigated. Literature on silver species (AgNPs or Ag^+) versus nucleic acids provided us with a clear picture that the damage to nucleic acids may depend on the concentration of silver ion. High dosage of silver ion (10 mg/L for several hours) could cause severe DNA molecules' condensation (Feng et al. 2000) while low dosage of silver species (0.1–1 mg/L of AgNPs, concentration of Ag^+ released must be even lower) could not lead to observable genome damage (Hwang et al. 2008; Sintubin et al. 2011). This dosage-dependent phenomenon explained quite well our results: lower dosage of AgNPs induced less inactivation (<3 log), where little genome damage occurred (<0.5 log). By contrast, higher dosage of AgNPs induced more inactivation (~4 log), where greater genome damage (>1 log) was observed (Fig. 4.14).

4.4 Continuous Flow System with Nanosilver

4.4.1 Continuous Flow System to Remove E. Coli

Applying AgNPs into practice is the ultimate goal of this study. In order to achieve this, continuous flow system incorporated with AgNPs was evaluated on inactivation of *E. coli* (Fig. 4.15). In the control experiment with the uncoated glasswool, the log inactivation of *E. coli* was nearly zero, indicating the glasswool used was unable to absorb or inactivate the bacteria. Any inactivation of *E. coli* was attributed to the AgNPs coated on glasswool. This served as the basis of this part of study.

Five milligrams or 10 mg of AgNPs were loaded on the glasswool and log removal of *E. coli* was observed to drop gradually in the first one-hour operation. Log inactivation by 5 mg of AgNPs dropped from 2.2 log (during startup) to 1.5 log (at the end of 1 h) while corresponding log inactivation by 10 mg dropped from 4.7 to 3.9 log (0–60 min in Fig. 4.15). Such performance of this system was

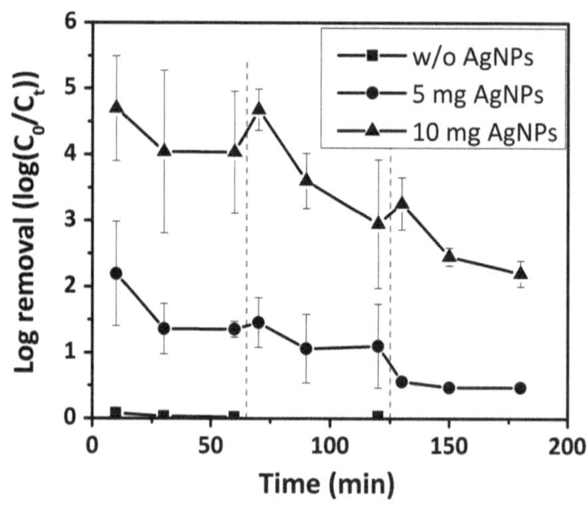

Fig. 4.15 Continuous flow system with AgNPs against *E. coli*. Dash lines denote backwash after 60 and 120 min, respectively

undesirable considering the feasibility and the sustainability of this system. It would be better if the performance could be recovered by some means.

Direct contact of bacteria to AgNPs was thought to be crucial in antibacterial effects of AgNPs. So one of the hypotheses for such efficiency decrease was that bacteria in the feed water was attached to the surface of AgNPs and accumulated there, leading to less active surface area of AgNPs available. Therefore, backwash by clean water was proposed and introduced after 1-h operation in order to wash away those dead bacteria [e.g. from 60 to 180 min (Fig. 4.15)].

The results of backwash shown in Fig. 4.15 verified this hypothesis. It clearly showed the antibacterial performance was recovered to some extent after backwash was adopted. In the case of 10 mg of AgNPs loading, the log inactivation of *E. coli* was recovered back to 4.5-log removal after first backwash at 60 min. The second backwash (120 min) was unable to improve the performance up to more than 4-log removal; but still, it increased by nearly 0.5 log after backwash. This study only evaluated the effect of backwash for the first two times. It was reasonable to believe that the backwash in the prolonged operation could still improve the performance to some extent, but the overall trend may remain decreasing.

Enhancement by backwash was, however, insignificant in the case of 5 mg loading. During backwash, it might be the case that not only dead bacteria were washed away, but small portion of AgNPs coated was leached away as well. Such loss might be negligible in case of higher loading (i.e. 10 mg) yet had an impact when only 5 mg of AgNPs was loaded. This might be the reason why backwash did not improve the performance in the 5-mg-AgNPs-loaded system.

In order to further exploit the potential of AgNPs in continuous flow system, air-assisted backwash (air scour) was adopted for the purpose of enhancing backwash performance. Air scouring can be used to increase water turbulence during backwash and thereby achieve satisfactory backwash performance at lower backwashing velocities. The reason for introducing air scour was same as aforementioned to further 'scrub' the bacteria off the surface of AgNPs. In this part of study, compressed air (100 mL/min) and water streams (40 mL/min) were pushed upwards simultaneously through the system. The air-to-water ratio was 2.5:1.

Comparison of air-assisted backwash and simple backwash is shown in Fig. 4.16. In the case of 10 mg loading, the air-assisted backwash achieved better performance than simple backwash by inactivating more than 5 log of *E. coli*. This improvement was also observed in the second backwash (120 min) and was statistically significant ($P < 0.05$). In addition, only 2 times of backwash with air scour were evaluated in this study. It was imaginable that air scour in the prolonged operation (10 mg of AgNPs loading) may still help recover the disinfection performance to some extent.

On the other hand, air scouring was not capable of recovering the 5-mg-loaded system back to more than 2-log removal. After the first air-assisted backwash, the log removal was even lower than non-air-assisted backwash. Again, one possible reason might be the loss of AgNPs during backwash and air scour might worsen such loss.

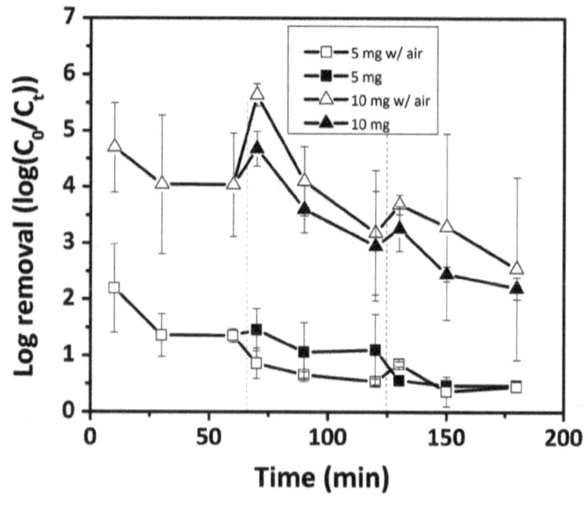

Fig. 4.16 Comparison the effects of air scour on the BW. *Dash lines* denote backwash after 60 and 120 min, respectively

4.4.2 Continuous Flow System to Remove MS2 Coliphage

Continuous flow system with AgNPs coated was also tested in terms of antiviral effects of MS2 coliphage. Figure 4.17 shows the preliminary results of 5 and 10 mg of AgNPs loadings, respectively. Similar to previous observation, glasswool column without AgNPs was unable to remove MS2 coliphage: basically the log removal by blank control was less than 0.2 log.

With the presence of AgNPs in the glasswool column, the performance of this flow-through system was not ideal regardless of either 5 mg or 10 mg AgNPs loading. The log inactivation rapidly decreased from 3.5 log at 10 min to less than

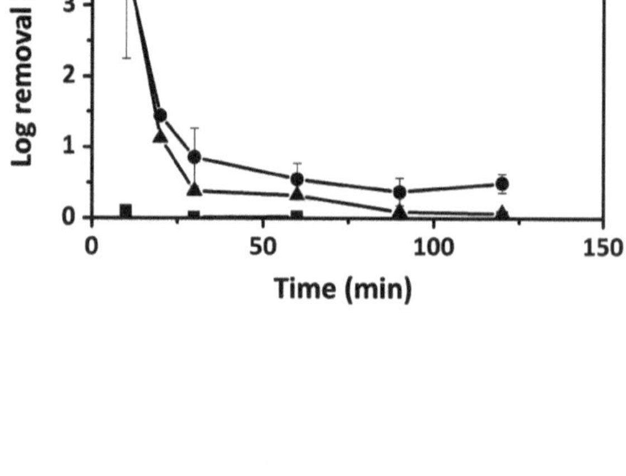

Fig. 4.17 Continuous flow system with AgNPs against MS2 bacteriophage

1 log removal at 30 min, and then stabilized at approximate 0.5-log removal till the end of experiment (120 min).

Unlike the flow-through experiment against *E. coli*, increase of AgNPs loading from 5 to 10 mg did not improve the disinfection performance to a large extent. This observation may again confirm that MS2 coliphage was more resistant than *E. coli* as what bench-scale test showed.

Studies on AgNP-involved continuous flow system against viruses are sparse. De Gusseme and co-workers evaluated removal of UZ1 bacteriophage in their AgNP-coated filter and approximate 3-log removal was achieved for 20 min (De Gusseme et al. 2009). Considering the resistance difference between UZ1 and MS2, the results shown here was comparable to theirs.

The latest study from the same group may provide us with hints on how to improve the antiviral performance. Approximate 2500 mg/m^2 of AgNPs was incorporated in PVDF membrane. This AgNP-contained membrane achieved 3.4-log removal for totally 225 L/m^2. Increasing dosage of AgNPs and better coating strategies are of the top priority in the future study.

References

De Gusseme B, Sintubin L, Baert L, Thibo E, Hennebel T, Vermeulen G, Uyttendaele M, Verstraete W, Boon N (2009) Biogenic silver for disinfection of water contaminated with viruses. Appl Environ Microbiol 76:1082–1087

Elechiguerra JL, Burt JL, Morones JR, Camacho-Bragado A, Gao X, Lara HH, Yacaman MJ (2005) Interaction of silver nanoparticles with HIV-1. J Nanobiotechnol 3:6

Fabrega J, Fawcett SR, Renshaw JC, Lead JR (2009) Silver nanoparticle impact on bacterial growth: effect of pH, concentration, and organic matter. Environ Sci Technol 43:7285–7290

Feng QL, Wu J, Chen GQ, Cui FZ, Kim TN, Kim JO (2000) A mechanistic study of the antibacterial effect of silver ions on *Escherichia coli* and *Staphylococcus aureus*. J Biomed Mater Res 52:662–668

Floyd R, Sharp DG (1978) Viral aggregation—effects of salts on aggregation of poliovirus and reovirus at low-Ph. Appl Environ Microbiol 35:1084–1094

Ho C-M, Yau SK-W, Lok C-N, So M-H, Che C-M (2010) Oxidative dissolution of silver nanoparticles by biologically relevant oxidants: a kinetic and mechanistic study. Chem Asian J 5:285–293

Hwang ET, Lee JH, Chae YJ, Kim YS, Kim BC, Sang B-I, Gu MB (2008) Analysis of the toxic mode of action of silver nanoparticles using stress-specific bioluminescent bacteria. Small 4:746–750

Jin X, Li MH, Wang JW, Marambio-Jones C, Peng FB, Huang XF, Damoiseaux R, Hoek EMV (2010) High-throughput screening of silver nanoparticle stability and bacterial inactivation in aquatic media: influence of specific ions. Environ Sci Technol 44:7321–7328

Kim JY, Lee C, Love DC, Sedlak DL, Yoon J, Nelson KL (2011) Inactivation of MS2 coliphage by ferrous ion and zero-valent iron nanoparticles. Environ Sci Technol 45:6978–6984

Kohn T, Nelson KL (2007) Sunlight-mediated inactivation of MS2 coliphage via exogenous singlet oxygen produced by sensitizers in natural waters. Environ Sci Technol 41:192–197

Levard C, Michel FM, Wang Y, Choi Y, Eng P, Brown GE Jr (2011) Probing Ag nanoparticle surface oxidation in contact with (in)organics: an X-ray scattering and fluorescence yield approach. J Synchrotron Radiat 18:871–878

Lok C-N, Ho C-M, Chen R, He Q-Y, Yu W-Y, Sun H, Tam PK-H, Chiu J-F, Che C-M (2007) Silver nanoparticles: partial oxidation and antibacterial activities. J Biol Inorg Chem 12:527–534

Matsumura Y, Yoshikata K, Kunisaki S, Tsuchido T (2003) Mode of bactericidal action of silver zeolite and its comparison with that of silver nitrate. Appl Environ Microbiol 69:4278–4281

Mattle MJ, Crouzy B, Brennecke M, Wigginton KR, Perona P, Kohn T (2011) Impact of virus aggregation on inactivation by peracetic acid and implications for other disinfectants. Environ Sci Technol 45:7710–7717

Morones JR, Elechiguerra JL, Camacho A, Holt K, Kouri JB, Ramírez JT, Yacaman MJ (2005) The bactericidal effect of silver nanoparticles. Nanotechnology 16:2346–2353

Pal S, Tak YK, Song JM (2007) Does the antibacterial activity of silver nanoparticles depend on the shape of the nanoparticle? a study of the gram-negative bacterium *Escherichia coli*. Appl Environ Microbiol 73:1712–1720

Park H-J, Kim JY, Kim J, Lee J-H, Hahn J-S, Gu MB, Yoon J (2009) Silver-ion-mediated reactive oxygen species generation affecting bactericidal activity. Water Res 43:1027–1032

Ratte HT (1999) Bioaccumulation and toxicity of silver compounds: a review. Environ Toxicol Chem 18:89–108

Romero OC, Straub AP, Kohn T, Nguyen TH (2011) Role of temperature and Suwannee River natural organic matter on inactivation kinetics of rotavirus and bacteriophage MS2 by solar irradiation. Environ Sci Technol 45:10385–10393

Sintubin L, De Gusseme B, Van der Meeren P, Pycke BFG, Verstraete W, Boon N (2011) The antibacterial activity of biogenic silver and its mode of action. Appl Microbiol Biot 91:153–162

Sintubin L, De Windt W, Dick J, Mast J, van der Ha D, Verstraete W, Boon N (2009) Lactic acid bacteria as reducing and capping agent for the fast and efficient production of silver nanoparticles. Appl Microbiol Biot 84:741–749

Sondi I, Salopek-Sondi B (2004) Silver nanoparticles as antimicrobial agent: a case study on *E. coli* as a model for Gram-negative bacteria. J Colloid Interface Sci 275:177–182

Yin L, Cheng Y, Espinasse B, Colman BP, Auffan M, Wiesner M, Rose J, Liu J, Bernhardt ES (2011) More than the Ions: the effects of silver nanoparticles on Lolium multiflorum. Environ Sci Technol 45:2360–2367

Chapter 5
Summary, Conclusions and Recommendations

5.1 Summary

This goal of this study was to systematically evaluate the feasibility of applying AgNPs as an alternative disinfectant in drinking water treatment process. This was achieved by accomplishing 4 sub-objectives, namely (1) to synthesize and compare different AgNPs; (2) to investigate the effects of relevant water matrices on AgNPs disinfection; (3) to study the antiviral mechanisms of AgNPs; and (4) to assess the performance of AgNPs in continuous flow mode. Results showed that bio-AgNPs was a promising nanosilver material that had advantage of small size and excellent stability, and was effective in inactivation against *E. coli* and MS2 coliphage in bench-scale tests.

TOC and chloride were two detrimental factors that can be toxic to antiviral activity of bio-AgNPs. Application of bio-AgNPs in real water sample should take these two parameters into consideration and proper pre-treatment may be necessary.

Antiviral mechanism study suggested that direct contact played the most important role in antiviral effects of bio-AgNPs. And such contact would induce significant capsid damage to MS2 bacteriophage. RNA damage of MS2 was only secondary.

Continuous flow system with 10 mg of AgNPs loading was found to be excellent in removing *E. coli* and backwash, especially air-assisted backwash was able to recover the performance to some extent.

To sum up, this study provided instructive information on selecting the best AgNPs (bio-AgNPs) and the suitable water quality (low TOC and low Cl^-) for future work in this field. In addition, study on antiviral mechanisms of AgNPs showed the significance of direct contact between AgNPs and viruses. It was implied that any practice to enhance such direct contact is worth a try to achieve better disinfection performance. Last but not least, continuous flow system with

X. Yang, *A Study on Antimicrobial Effects of Nanosilver for Drinking Water Disinfection*, Springer Theses, DOI 10.1007/978-981-10-2902-8_5

AgNPs was demonstrated to be feasible and promising. Engineering solution (backwash with air scour) could be useful in recovering the disinfection performance and more detailed research could be done in the future.

5.2 Conclusions

Nanosilver from three sources, namely com-AgNPs, chem-AgNPs and bio-AgNPs, were successfully obtained either from industry or from synthesis in the lab. TEM images confirmed the presence of nanoscale particles; however, it was found by TEM at the same time that both com-AgNPs and chem-AgNPs had aggregation to varying degrees. Zeta potential results of these three materials showed they were all negatively charged. And HDD confirmed the aggregation status of com-AgNPs (322 nm) and chem-AgNPs (122 nm). Histogram of size distribution of bio-AgNPs was obtained by multiple TEM images, further verifying that bio-AgNPs had the smallest particle size (20–40 nm) among the three.

These three AgNP materials were subsequently compared in terms of log inactivation against *E. coli* and MS2 coliphage. Bio-AgNPs, which was with smaller particle size and less aggregation than the other two, achieved the better performance against both *E. coli* and MS2 coliphage. Ten milligrams per litre of bio-AgNPs was capable of inactivating all *E. coli* (>6.3 log) and achieving 4-log MS2 inactivation in 5 min. Bio-AgNPs disinfection against MS2 was optimized and 10 mg/L for 10-min contact time in bench-scale study was the recommended operating condition if 4-log viral removal is required.

Effects of water matrices, such as ionic strength, hardness, TOC, chloride and pH, were subsequently evaluated with bio-AgNPs at a concentration of 10 mg/L for a contact time of 10 min. Ionic strength and hardness would only have moderate effects on the disinfection performance. High ionic strength (5.6 mM) and hardness (188 mg/L as $CaCO_3$) had similar impacts and could only achieve approximate 1.5-log removal of MS2. TOC and chloride were found to affect significantly the viral inactivation of bio-AgNPs, in which 30 mg/L of TOC (as humic acid) and 5 mg/L of Cl^- fully quenched the antiviral effects of bio-AgNPs. If we intend to apply nanosilver in drinking water disinfection process, a pre-treatment is requisite to meet some standards: TOC should be less than 1 mg/L and chloride less than 1 mg/L in order to achieve 4-log removal of virus.

Systematic mechanism study of antiviral effects of nanosilver was conducted in terms of both mode of action and viral damage. Direct physical contact accounted for the majority of viral inactivation of MS2 coliphage. Silver ion released and ROS production, however, were found to contribute little to antiviral activities against MS2 coliphage. This finding was seemingly contradictory to previous study (Sintubin et al. 2011) but still reasonable due to the resistance difference to disinfectants between bacteria and viruses. As for viral damage, ELISA assay results showed significant decrease of antigenicity of MS2 coliphage, suggesting substantial capsid damage of MS2 after bio-AgNPs treatment. Quantitative RT-PCR demonstrated only minor genome damage occurred following the AgNPs treatment. This finding further translated into that it was less likely for AgNPs to induce genome damage.

Continuous flow system with AgNPs coated was demonstrated to be effective in inactivation of *E. coli* at 10 mg of AgNPs loading. Backwash of the AgNPs-coated glasswool column was capable of recovering the disinfection capacity against *E. coli.* by approximate 0.5 log. Furthermore air-assisted backwash was also tested and found to be a significantly greater ($P < 0.05$) improvement in recovering the inactivation of *E. coli.* On the other hand, continuous flow system with AgNPs was unable to remove MS2 coliphage effectively. The performance decayed fast from 2.5 log at 10 min to less than 1 log at 30 min. Increasing AgNPs loading from 5 mg to 10 mg did not improve the performance.

5.3 Recommendations

Based on the findings in this study, recommendations for work in the future are listed as follows:

- Bio-AgNPs are not the end of this topic. Further study could be done in synthesizing AgNPs of smaller size, improved stability and greater ease of synthesis. If it could be achieved, the performance of AgNPs could be enhanced beyond what was observed in this study and become more promising to be developed as an alternative disinfectant.
- Based on our finding that TOC and chloride were toxic to AgNPs, water quality of feed water for AgNPs treatment should be strictly controlled. Feed water with little TOC and chloride would be preferred. Specific scenario, such as rainwater disinfection, can be considered.
- Mechanism study can be more detailed. When studying the contribution of Ag^+ release, the concentration of silver ions may be higher in surrounding area of AgNPs than that in bulk solution. Identifying and quantifying the silver ions near the surface of AgNPs could be a milestone towards better understanding the killing mechanism at molecular level.
- Conditions of backwash and air scour, e.g. backwash flowrate and air-to-water ratio, need to be optimized to achieve the best recovery. Rapid decrease in antiviral effects of continuous flow system with AgNP shall be addressed by future research focusing on both backwash optimization and better coating techniques.

Reference

Sintubin L, De Gusseme B, Van der Meeren P, Pycke BFG, Verstraete W, Boon N (2011) The antibacterial activity of biogenic silver and its mode of action. Appl Microbiol Biot 91:153–162

Appendix

See Fig. A.1.

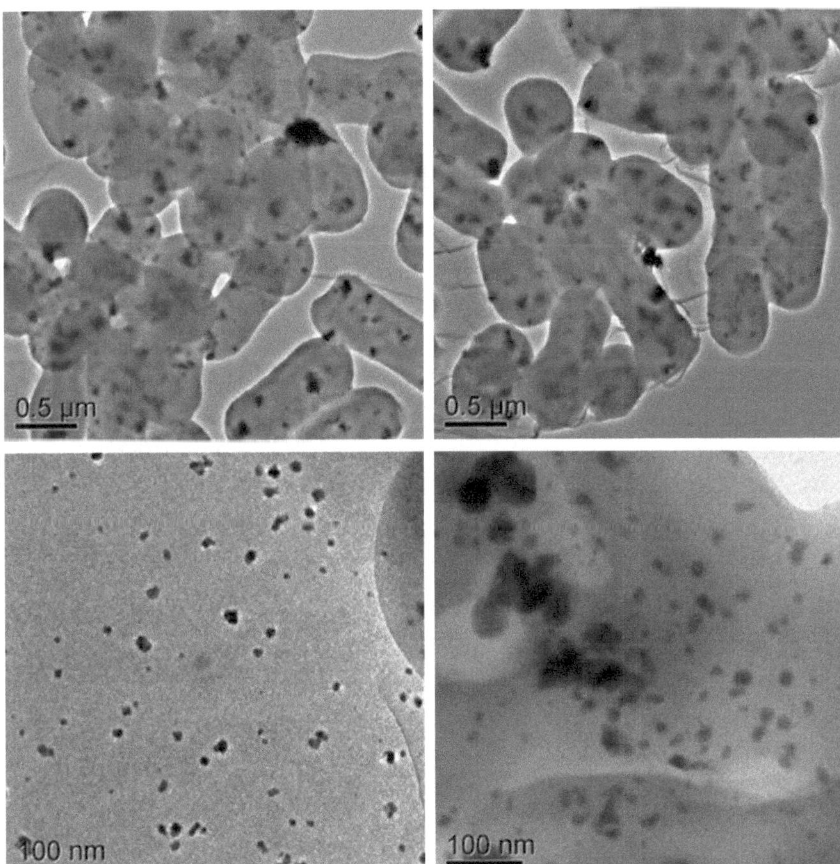

Fig. A.1 TEM images of bio-AgNPs used for size distribution

X. Yang, *A Study on Antimicrobial Effects of Nanosilver for Drinking Water Disinfection*, Springer Theses, DOI 10.1007/978-981-10-2902-8